Organic Chemistry

an informal history

Paperback Edition 2019

George E. Parris, PhD

Preface

I'm and organic chemist and my career has spanned over one-third of the period focused on in this book. I knew I wanted to be a chemist in when I was 15 years old, and after my exposure to organic chemistry in college (1966) I was hooked. A large part of that had to do with my association with G. Gilbert Long, George O. Doak and Leon D. Friedman at North Carolina State University. This was re-enforced by Carl L. Bumgardner and William P. Tucker. Organic chemists (for the most part) seemed to be *outgoing* and *pragmatic*; yet, in many ways, *artistic*. And, they definitely were not chemophobic. The sights, sounds and smells of chemicals seem to raise their enthusiasm. I find organic chemistry to be very visual and intuitive. I love to "see" the molecules and watch the bonds move. I love to make things. *Making things* (i.e., synthesis) has played a large role in organic chemistry. *Measuring things* is more in the domain of the physical and analytical chemists. Inorganic chemists have been somewhere in the middle. However, I must confess that the inclusion of physical measurements (e.g., kinetics, equilibria and spectroscopy) into organic chemistry has definitely enriched my experiences.

My first taste of organic chemistry came in 1962 when my father brought home two books by Isaac Asimov: *The World of Carbon* and *The World of Nitrogen*. By the time I had finished Rakoff and Rose (1966), Roberts and Caserio (1964) and March (1968), I was ready for graduate school. But my most important education came when I got a job as a lab assistant to G.G. Long in the spring of 1966. For the next three years I worked full time in his laboratory during the summers and part time during school. For the most part, I made organic compounds of arsenic and antimony (e.g., trimethylstibine, trimethylantimony dichloride, tetramethylstibonium salts, 1,4-bis(diphenylarsyl)butane, etc. I worked with chlorine, bromine, sodium in liquid ammonia, trimethyloxonium tetrafluoroborate, Grignard reagents, methyl lithium, all the common acids and bases, and liters of diethyl ether, benzene, carbon tetrachloride, chloroform, methylene chloride, acetonitrile, methanol, ethanol and other solvents (including gallons of acetone used to clean glassware). I created more hazardous waste in an afternoon than most undergraduate chemistry departments make in a year these days.

But, through care and caution, I never had a serious accident at NC State or as a graduate student. I did have an embarrassing excursion, the first time I made a mole of trimethylstibine. The reaction involves making a methyl Grignard and then reacting it with antimony trichloride. I did it on about three-liter scale. Then, the ether and trimethylstibine (b.p. 81ºC) are distilled and chlorinated to make trimethylantimony dichloride, which recrystallizes from acetone to give amazing monoclinic crystals. But the reaction pot now containing a ball of magnesium salts, apparently partially hydrated with ether, has to be cleaned up and put away. Thinking nothing of it, I quickly added about 100 mL of water and was greeted with a very exothermal reaction (solvation of the magnesium chloride I assume) and volcanic eruption of boiling water and ether. It was a big mess in the hood and I recall some splashed on my face and hands…but no pain and no skin damage.

Nonetheless, while I was an undergraduate, there were two serious fires in Dr. Long's lab. One involved a first-year graduate student who was trying to evaporate several liters of ether from some reaction mixture. Rather than use the rotary evaporator, he had the ether in a 4-liter filter flask; and he decided to put a large rubber stopper in the top and draw the ether vapor off with the water aspirator. This worked, but the evaporating ether soon cooled the flask. He had the bright idea of putting it on a hot plate/stirrer. Then he left the room. He returned just as the ether pressure in the filter flask popped the stopper and sent a column of vapor to the ceiling. The vapor settled back on the hot plate and flashed from bottom to top. I did not see the incident and only learned about it the next morning when I came to work and found a foot of foam blocking the door to the lab. In another episode, another undergraduate (with a job like mine) came into the lab in the evening and was showing a friend of his around. A one-pound cylinder of sodium metal sat on the shelf in a gallon pickle-jar of kerosene. The kerosene made the jar slippery, and immediately below the jar was a water bath. I think you can guess what happened next. When I came in the next day, the lab was like a blackbody. Soot (from the burning kerosene) had completely covered everything: The windows, the lights, the benches and the glassware. The undergraduate spent the next month washing glassware.

Also, as an undergraduate at NC State (in old Withers Hall), there was one day a rumble on the third floor. At the end of one of the teaching labs, a concrete block wall had been erected to create a small graduate student laboratory about 10 feet wide with one hood. Some graduate student was preparing diazomethane and had the flask fall out of the hood onto the floor. The explosion was not all that large…the student was unhurt and windows were not blown out although the blinds were shredded; but the pressure wave pushed over the concrete block wall, which then fell on the department's C/H/O analyzer, which sat beside the wall. In these three episodes no one was hurt. But that was not the case at Georgia Tech.

As a senior, Dr. Long arranged for me to present some of my work at the South Eastern Regional ACS meeting in Tallassee, FL. At the meeting, we had dinner with Eugene C. (Gene) Ashby who was a charismatic guy doing organometallic chemistry at Georgia Tech. He invited me to apply. But I sort of had my heart set on what I considered to be the big leagues…MIT. I arranged to make a trip to visit Dietmar Seyferth. It must have been February or March 1969. The Vietnam War was in full swing.

The airplane from Raleigh-Durham seemed to nosedive into Boston's Logan airport; both of my ears plugged up and the right one would not clear. By the time I was on the ground, I was in deep pain and I could barely hear. Looking out the window of the airplane as it taxied, I noticed that the wings barely cleared piles of snow. Somehow, I got out of the airport and onto the Boston subway. Unfortunately, I made the mistake of attempting to use the restroom in the subway. My impression of the restroom was that people (apparently bums) must have stood just inside the door and peed in the general direction of the toilet…Disgusting. Once in Cambridge, I trudged along icy walkways with piles of snow three-feet high on either side. Finally, I reached Dr. Seyferth's office where I met a surly administrative assistant in an outer office, who only grudgingly allowed me access to Seyferth's inner sanctum.

I only remember one thing about the conversation with Seyferth: There was not a word about chemistry, but he emphatically told me that MIT could not protect me from the draft…something that I had never even considered. In the tour of the building, I noted the labs were older than Withers Hall at NC State (State was about to open Dabney Hall for Chemistry) and the NMR room I was shown looked like a junk pile with a sad 60 MHz instrument sitting there (at NC State, there was a full-time technician to run samples on a nice 100 MHz instrument). Finally, when the classes changed, it seemed that the chemistry department shared space with some liberal arts groups and the halls were flooded with bearded students adorned with various anti-war symbols. I had no regrets when I left; I was eager to return home.

The next day, I called Gene Ashby and told him I wanted to come to Atlanta (sight unseen) and I arrived there near the end of June 1969. The Georgia Tech labs were of the same vintage as MIT, but much more organized than MIT (at least the part that Seyferth had access to). At GIT, there were 60 and 100 MHz NMRs with technical assistance and a premier glass shop run by Don Lilly. And I believe that Ashby had assembled the premier organometallic/metal hydride group in the world. They had about 6 functional inert atmosphere boxes, a full-scale vacuum line (with gallons of liquid mercury), two GCs, IR, and AA equipment in the group.

The Ashby group of about 15 students and post-docs was divided more or less into two groups: one doing organic chemistry of Grignard and organo-aluminum reagents (composition and reactivity) and one doing metal hydride synthesis and characterization (rocket fuel and reducing reagents) chemistry. I fell into the organic reagent composition area.

Shortly before I arrived, one grad student lost a finger (literally, it was never found) in a dry box explosion (the hypothesis was that potassium superoxide formed from excess potassium metal left over from synthesis of (and comingled with) potassium hydride). But my work went safely, easily and successfully during the first year. That year, my lab mate Robert C. (Bob) Arnott astounded me by dissolving potassium hydride in a benzene solution of (as I recall) di-isobutylmagnesium to make the $K(MgR_2H)$ complex. When I could not turn down any more of Bob's invitations, he got me into the world of spelunking (cavers call it "caving") at a place called Cemetery Pit near Rising Fawn, GA.

Around Christmas 1969, there was a draft lottery. Sure enough, I was drafted and entered service (September 1970) shortly after I was married (June 1970), and spent the next 18 months (until April 1972) working for Uncle Sam. The Vietnam War was winding down and every draftee was getting an early out. My turn came in April 1972. I returned to Georgia Tech with a Fannie and John Hertz Fellowship and my wife in May of 1972 and worked successfully on several projects and officially

graduated in December 1974.[1] My relationship with
Ashby was strained. Although he and I were
politically on the same side, I did not like his research
ideas.

Pretty much without Ashby's approval, I went off and
tried to make a known compound dicyclopentadienyl
magnesium, because it could be sublimed. I prepared
some cyclopentadiene from the dimer and added two
moles to a mole of dimethylmagnesium in diethyl ether
assuming that methane would be released. Nothing
much happened. There were a few tiny bubbles but I
thought that might be moisture in the cyclopentadiene.
I set the reaction flask under a bubbler to allow any gas
to be release without exposure to air and left for the
weekend. When I returned, I could not believe my
luck, the flask was half-full of transparent needles that
turned out to be a di-etherate of

---

[1] To my delight, two of the high-profile MIT professors had left:
FA Cotton went to Texas A&M and HO House had come to
Georgia Tech and was on my thesis committee! Moreover, while
I was on active duty, the department moved into a new
chemistry building (i.e., I did not waste my time moving dry
boxes).

bis(cyclopentadienylmethylmagnesium); i.e., a methyl-bridged dimer.[2]

Ashby started nagging me to take on a project that another student had been working on for a couple of years with limited success: alkylmagnesium alkoxides. From what I had heard, it sounded like an intractable mess; and I really did not want to get bogged down with it. But I had about worn out my good will with Ashby, so I agreed to take on the project. I started from scratch…never read a word of what the other student had done. It tuned out that the chemistry was actually very well behaved…the key recognition was that there was a very slow rearrangement of methylmagnesium t-butoxide occurring in ether. When I realized that, everything fell into place. In relatively non-polar diethyl ether, the initially-formed dimer slowly reorganized into a tetramer (cubane of Mg-O bonds). This tetramer was soluble in benzene, so I decided to

---

[2] Because this is soluble in nonpolar solvents and the ether groups are readily dissociated, it might be interesting to see what it would do with olefins. It looks a lot like a Ziegler–Natta catalyst. G.E. Parris and E.C. Ashby. 1974. The composition of Grignard compounds. IX. The structure and solution composition of cyclopentadienylmethylmagnesium in benzene and ether solvents. *J. Organomet. Chem.* 72:1.

try to sublime it. Sure enough, it sublimed un-changed. This was a novel outcome in 1974. I'm sure it would be an interesting alkylating agent and it would be interesting to study the kinetics of the rearrangements.

But I was done. I just wanted safely out of graduate school. I had pissed Ashby off (justifiably) by submitting a letter to the editor of *Inorganic Chemistry* describing the effects of solvation on the Schlenk equilibrium. I did that unilaterally because Ashby had scoffed at my idea in a group meeting. When the letter got accepted, I broke the news to Ashby (in the context that I had published several papers without his involvement, as an undergraduate and while in the Army). He was not pleased and I agreed to withdraw the paper, which I did. Thus, when I wrote up the methymagnesium alkoxide work[3], I listed the authors in the order that they had been interested in the work...Ashby first and me last. Of course, as organic chemists, we had no interest in magnesium oxide. But it turned out later that this compound is used to

---

[3] E.C. Ashby, J.A. Nackashi, and G.E. Parris. 1975. The composition of Grignard compounds. X. NMR, IR, and molecular association studies of some methylmagnesium alkoxides in diethyl ether, tetrahydrofuran and benzene. *J. Am. Chem. Soc.* 97:3162.

deposit magnesium oxide from the gas phase onto microelectronics (by thermal decomposition yielding methane and isobutylene). The crystal structure determination also confirmed my cubane proposal. My fellowship would have allowed another year at Tech, my wife and I doing very well financially, but the job market had evaporated for PhD chemists and I felt that Ashby had lost his edge, giving up chemistry for religion.[4] So, I walked away from some interesting chemistry.

Near the end of my tenure in Atlanta (about September 1974) while I was finishing up my thesis, a new graduate student had started working, apparently with little or no supervision. For reasons I do not know, he had tried to react sodium with hexachlorocyclopentadiene on a large scale. I came

---

[4] I think that Ashby could have won a Nobel Prize had he shown the interest that drew me to him originally, but somewhere in the early 1970s he began expressing a much more fundamentalist religious interest. The war and the Watergate affair created substantial friction among various group members. There were loud, daily arguments, which I tried to avoid. I felt the ship was sinking and took the first opportunity out. Professor Ashby has retired and written books that can be described as defense of "old-earth creationism/intelligent design."

into the building one afternoon with my infant son[5] in a backpack and found blood on the floor in the corridor and a mess of blood and soot in the lab where the new student had been working. My colleagues explained that he had had an explosion and fire and he had been taken away by ambulance. Over the next couple of days Ashby was noticeably shaken. I visited Atlanta in 1977 and came by the lab. I met the guy (really for the first time) …a ragged beard was growing over his scared face and both of his arms were covered in canvas sleeves that hooked around his thumbs.

I left Georgia Tech about October of 1974 and took an NRC post-doc with the Inorganic Materials Division of the National Bureau of Standards (now the National Institutes of Standards and Technology). I got the position by proposing to look at the environmental chemistry of arsenic and antimony and I published a couple of papers with Fred Brinckman on quaternization (Sn2) and oxidation of trimethyl-arsine and –stibine.

It was a two-year appointment and I was eagerly looking for an academic position. The only interview I ever got was at Auburn University. I don't recall even

---

[5] I never took a small child into a chemistry building again.

getting a letter telling me who they hired. With unemployment looming, my mentors at NBS (NIST) had a bright idea. The US Environmental Protection Agency (then at 401 M St SW, DC) was gearing up for the Toxic Substances Control Act, which was described as having lots of "testing requirements." I was sent as the object of an interagency agreement to the Office of Toxic Substances (1975). This arrangement soon proved untenable and I left the laboratory entirely for the EPA full-time job.

The Toxic Substances Control Act was passed in September 1976.   It did not take me long to realize that the EPA had a lot more to do with politics than science. And after one annoying episode, I absconded to the Industrial Chemistry Branch (Bureau of Foods) of the Food and Drug Administration (FDA), where I had my most serious personal accident (200 C St. SW Washington, DC) a stone's throw from the US Capitol (1978). I was interested in protein adducts of xenobiotics and I wanted to make a methylating agent by dissolving HCl gas in methanol. Unfortunately, we had a lecture bottle of HCl gas, but I could not find the relevant needle value fitting. The bottle was tiny compared to the chlorine cylinders I had worked with and I was not even sure that it had any gas in it. Thus,

I jury-rigged plastic and rubber tubing together thinking "how big a deal can this be." The whole set up was, of course, in a hood and I was very cautious…but not prepared for what was about to happen. Carefully, very carefully, I opened the value on the cylinder. To my dismay the gas pressure completely blew apart the jury-rigged connections and I was engulfed in HCl gas.[6] I could feel it in my eyes, in my nose, mouth, even upper lungs and of course my face and hands. I consider myself to be fairly cool under pressure, and remember I had my hand on the valve. But I did not have the presence of mind to turn the valve off (I had never contemplated it). But I did have the presence of mind to bolt to the safety shower, which did work. My punishment was to stand in the corridor of a government building under a shower for 15 minutes. But, by the end of the day, I had dried out and took the metro home. My colleagues got the hood sash down and that little lecture bottle continued to blow HCl, which formed a cloud with humidity in the

---

[6] I did not consider it then, but the vapor pressure of HCl (l) at 20°C is over 40 atm! If I had cooled it in dry ice acetone (-78°C), the pressure would have been about 2 atm and I might have gotten away with this. It is hard to appreciate the affect that dissolution in water has on reducing the vapor pressure of HCl and ammonia.

air before it was exhausted from the hood, for over 10 minutes.

After three years, I became bored with the government and frustrated with the glacial movement of document review. I went to work for a consulting firm and spend the next 20 years making my living as an expert on government regulations. In 2003, as a result of an unexpected changes in regulatory standards, the trade association that I ended up working for went out of business leaving me as a 55-year old unemployed. Since that time, I have been looking for a full-time job in chemistry and supporting my family as an adjunct professor at a variety of colleges and universities, while making ends meet by selling cars. Along the way, I became interested in a number of biological issues and largely shifted my focus from pure chemistry to applied chemistry, biology and material science.

Scope

This book will cover an exciting 150 years. I have broken the book into segments that I have identified as follows:

Prelude

Empirical (1820-1870)

Classical (1870-1930)

Early Modern (1930-1955)

Modern (1955-1980)

Post-Modern (1980-2020)

To make this task manageable, I have left out biochemistry and molecular biology. I have included short sections on major instrumentation that facilitated revolutionary progress in various areas (i.e., bomb calorimetry, infrared spectroscopy, column chromatography, gas chromatography, mass spectrometry, nuclear magnetic resonance, high performance liquid chromatography, etc.). I have tried to mention all the chemists and principal chemical reactions that you are likely to come across in an introductory (two-semester) course in organic chemistry and some material from advanced courses. I have also attempted to address economically important chemical processes, which are not too exciting to organic chemists, but which are incredibly important to the average person: plastics, fibers and fuel. Some may regard my attention to the petroleum industry (especially evolution of gasoline) to be excessive; but as I assembled this book and though about the heroes of modern organic chemistry, I was caused to ponder the economic importance of, for example, the total synthesis of B_{12} versus the invention of tetraethyl lead. What organic chemists find exciting is not necessarily what makes organic chemistry move forward. In the last chapter (post-modern organic chemistry), I express my opinion that organic chemistry seems to be maturing and stagnating. The major new fields (e.g., pollution control, environmental remediation and "green chemistry") actually are signs of *retreat, not advance* for this field. Thus, I draw the controversial conclusion that organic chemistry has allowed itself to become sterile, in part, through academic inbreeding leading to a failure to embrace the fields now known as "molecular biology," which I describe *as organic chemistry invented by biologists* and materials science of carbon (i.e., *organic chemistry of pure carbon*).

George E. Parris, PhD

Gaithersburg, MD

August 2015

Paperback Edition, August 2019

I. Prelude

1. The Theory of Vital Force

The state of chemistry in the early 1800s included an emerging understanding of the elements that fully displaced alchemy, but a confused understanding of heat (and energy) and ambiguity concerning the existence of atoms and molecules. There were people and pockets of academia that fully understood heat as the kinetic energy of molecules in random motion, but the primary centers of science (England and France) in Western Europe still clung to the caloric theory of heat and were generally agnostic to the existence of molecules. I have already discussed this state of affairs at length in my book *Matter and Energy an informal history* (self-published on-line via Amazon/Kindle, April 2015).

Here, I want to focus more specifically on the branch of chemistry known as "organic chemistry," which owes it genesis to the notion believed by many and expressed most authoritatively by Jacob Berzelius

(1779–1848). The idea was that substances could be divided between mineral and non-mineral. Life was (and is) something mysterious and it was clear to most people that living systems had material properties that were someway related and distinct from the mineral substrate of earth. The concept became known as *vitalism*, which had certain religious overtones: Living things were unique and produced substance that were never available through non-living means. After discovering cerium (1803), Berzelius gained international recognition and by 1806 people listened respectfully when he proposed that organic materials could only be formed by living systems through *vital force*.

Of course, he was wrong and there was already evidence of the error. Louis Claude Cadet (1731–1799) had heated potassium acetate (obtained by reacting potash with vinegar) with arsenic(III) oxide (a byproduct of roasting copper ores and one of the preferred poisons of the middle ages and Renaissance) to make cacodyl in 1760.

$$4 \ KO_2CCH_3 + As_2O_3 \rightarrow (CH_3)_2As\text{-}O\text{-}As(CH_3)_2 + 2 \ K_2CO_3 + 2 \ CO_2$$

$(CH_3)_2As-O-As(CH_3)_2 + O_2/H_2O \rightarrow$ disproportionation products

Including Cacodyl $CH_3)_2As-As(CH_3)_2$

Cacodyl Oxide and Cacodylic Acid $(CH_3)_2AsO_2H)$

Cadet's fuming red liquid with the evil smell may not strike you as an "organic compound" but the biomethylation of arsenic[7], antimony and even bismuth are now well established and biological processes yielding cacodyl derivatives and arsenosugars are known.[8] [Arsenic seems to mimic nitrogen (not phosphorus) in biological systems.]

Regardless, science progressed. One of Berzelius's most productive students was Friedrich Wohler (1800–1882) who synthesized silver cyanate and determined its empirical formula to be AgCNO. Concurrently, Justus von Liebig (1803–1873) synthesized silver fulminate and determined its empirical formula also to be AgCNO. The difference was that Liebig's compound was detonated by shock and Wohler's was

---

[7] First appreciate by Bartolomeo Gosio (1863-1944) in 1892.

[8] R. Bentley and T.G. Chasteen. Microbial Methylation of Metalloids: Arsenic, Antimony, and Bismuth. *Microbiol. Mol. Biol. Rev.* 66(2):250-271 (2002).

not. These two met in 1826 and soon agreed that the products had the same formula, but very different properties. Berzelius soon extrapolated this idea to the concept of *isomers*.

Ironically, in 1828 Wohler heated ammonium cyanate (an inorganic material) and realized he had produced urea (which had previously only been isolated from urine)[9]. This observation had little notice. Based on the defense of the theory of vital force that Berzelius displayed with people outside his research group (see below) there is little wonder why Wohler did not push this point.

Berzelius died in 1848 and with him went the last of the theory of vital force. A roadblock to a modern (e.g., structural) interpretation of organic chemistry was also removed. However, among his followers, especially Hermann Kolbe (1818–1884), new roadblocks appeared.

---

[9] Attributed to Herman Boerhaave (1668–1738) who developed gout in 1722 cited by F. Kurzer and P. M. Sanderson Urea in the history of organic chemistry: Isolation from natural sources *J. Chem. Educ.* 33(9):452 (1956).

2. Organic Radical Theory

Liebig and Wohler continued their collaboration in Germany working with the oil of bitter almonds, a source of cyanide. Cyanide (CN-) had been known for some time to behave as a fundamental unit of organic and inorganic chemicals. It was regarded as a "root" part and was called a *radical* (i.e., "from the root"). Through a series of transformations, Liebig and Wohler found that other pieces of organic molecules (i.e., the benzoyl group) also were transferred as a unit (1832).[10] In the absence of a good atomic theory and with doubt about the existence of molecules, Berzelius took the view that these radicals were elementary building blocks of organic molecules (analogous to atoms of a metal)[11] and that they were (as in the case of salts) held

---

[10] Wohler and Liebig. Investigations of the radical of benzoic acid, *Annalen der Pharmacie*. 3:249-282 (1832). Benzaldehyde is a major component of almond extract and in the presence of cyanide it condenses to benzoin.

[11] Perhaps Berzelius hoped to save the idea of vital force by equating organic radicals with the (inorganic) elements that were being isolated. Radicals were regarded as "super atoms" until the structure of organic molecules was understood.

together exclusively by electrostatic forces (i.e., positive and negative attracting).[12] There was really no understanding of structural chemistry at that time that would allow any chemist to distinguish an organic radical from a cluster of elements (i.e., atoms). When Auguste Laurent (1807-1853)[13] presented evidence (1836) that chlorine (a "negative element") could be substituted for hydrogen (a "positive element") in chlorinated ethanol; and thus undermined Berzelius's idea of clusters of ions with opposite charges making up radicals, Berzelius attacked him viciously and effectively prevented Laurent from obtaining access to the more famous laboratories.

Ironically, Robert Bunsen (1811–1899) began studying cacodyl in 1838 and realized that families of compounds could be made from the $(CH_3)_2As$- radical. At the time, the dimers of radicals were typically mis-understood to be the radicals themselves. For example,

---

[12] Recall that at this time, salts were assumed to not dissociate in solution. Thus, it is easy to understand that Berzelius and later Kolbe believed organic radicals could be composed of ionic materials that did not dissociate in solution.

[13] A student of Jean Baptiste Dumas (1800–1884). Dumas worked with the chlorination of ethanol to chloral (trichloroacetaldehyde) and trichloroacetic acid.

cacodyl $(CH_3)_2As-As(CH_3)_2$ was assumed to be the radical $(CH_3)_2As-$; ethane was assumed to be methyl radical; and butane was assumed to be ethyl radical. This work helped to solidify the concept of "a radical" among organic chemists. A student of Bunsen and Wohler, Hermann Kolbe managed to produce acetic acid from carbon disulfide (1845). And this, coupled with the passing of Berzelius (in 1848), finally dismissed the theory of vital force.

Development of the Theory of Organic Radicals

Kolbe spend 1846-1848 in England in the laboratory of Lyon Playfair. He brought with him the techniques of gas analysis (i.e., applications of the gas laws with gravimetric measurements) and shared them with Edward Frankland (1825-1899)[14] who came to Playfair's laboratory about the same time. Together they followed up the work of Liebig and Wohler on nitriles by acid hydrolysis of several nitriles (R-CN) to the corresponding acids ($R-CO_2H$). This helped

---

[14] Frankland was a bastard who took his mother's surname. His biological father must have been a wealthy and reasonable man who paid his mother a generous annuity to maintain his anonymity.

established methy-, ethyl- and butyl- as radicals (i.e., pieces of the molecules that did not change).

Frankland's interest in the reactions of metals with organic radicals apparently came during the summer of 1847 when he accompanied Kolbe back to Bunsen's laboratory in Germany. In a puzzling experiment, he attempted to prepare ethyl radical by reaction of ethyl nitrile (CH_3CH_2-CN) with potassium. The reaction they observed was violent and undoubtedly produced a mixture of small hydrocarbon molecules, which they interpreted as methyl radical (i.e., ethane) contrary to the ethyl radical they had expected. Frankland probably realized that potassium was too reactive and caused unexpected products.

Thus, when Frankland returned to England (fall 1847) he adopted Bunsen's successful use of zinc, which had converted cacodylic acid into cacodyl (radical). With ethyl iodide and zinc filings sealed in a heavy glass tube, Frankland did not observe any reaction until he heated the tube to over 150°C. Ultimately, he observed reaction resulting in a crystalline solid and mobile liquid. But he was not able to analyze his products until he returned to Bunsen's lab in Germany (October

1848).[15] Analyses of the reaction products after quenching with water yielded a substantial volume of gas as the liquid (ethyl radical; i.e., butane) was volatile. The products were typically in a volume (i.e., mole) ratio ethylene 22%, ethyl radical (i.e., butane) 50%, methyl radical (i.e., ethane) 26%, with about 2.5% nitrogen. Frankland apparently ignored the oddity of forming methyl radical (ethane) from ethyl radical (ethyl iodide).[16] A modern interpretation suggest that the reaction of ethyl iodide with zinc at high temperature yielded primarily butane by some sort of radical coupling and the C_2 gas (ethane/ethene) resulted from extraction of a hydrogen from one ethyl group by another producing ethane:

$$\text{Ethylene + Et-H} \leftarrow \text{2 Et} \rightarrow \text{Et-Et}$$

Nonetheless, Frankland believed he had isolate ethyl radical and entitled his papers "On the isolation of organic radicals." The vapor densities reported by Frankland were soon argued by Laurent to indicate

---

[15] He continued these studies in the winter of 1848 and into the spring of 1849 with Bunsen, receiving his doctorate from Bunsen at the end of June 1849.

[16] When zinc, ethyl iodide *and* water were reacted under similar conditions, only "methyl radical" (ethane) was obtained.

that the empirical formula should be doubled. Perhaps Frankland sensed some problem with the radical theory as it was being advanced, because he shifted the focus of his work with zinc as we will see below.

Kolbe became the chief protagonist of the radical theory. Of course, the idea of the organic radical is very useful today: We just assign the part of the molecule that is not reacting to a radical (usually designated as R- or Ar-) and it is convenient to use this nomenclature when describing series of compounds. Although today, we view organic radicals as little more than a convenient nomenclature device, in Kolbe's day the idea of a "radical" conveyed an implication of some uniquely stable assembly of atoms.

Herman Kolbe's View of Organic Radicals

Interestingly, Herman Kolbe's name appears in this book more than any other chemist. He had a long career, published important work, mentored productive students, and ultimately made a fool of himself. This is rather sad, because he had a relevant point to make. I want to make this point here in some detail to avoid discussing it each time it arises:

At the time that Kolbe worked (1850-1880), indeed up until the work of Schrodinger and Pauling in the 1930s, the only physical force that chemists understood to influence atoms was electrostatic interactions, which had been established by Charles-Augustin de Coulomb (1736–1806) about 1785. Magnetic interactions were also understood, but the magnetic moments of electrons were not known until 1925 (George Uhlenbeck and Samuel Goudsmit). Indeed, the very existence of atoms and molecules was not fully accepted until the early 1900s. Kolbe was convinced of the existence of atoms and ions (atoms with charges) and when it became apparent that certain groups of atoms clustered together (i.e., organic radicals), he accepted that. But, recall that electrostatics do not have any preferred orientation.

Thus, with no forces other than electrostatics to hold atoms and ions together, the only model of organic radicals that had actual theoretical support from 1832 to 1932 was the idea that organic radicals were nothing more than unusually stable clumps of ions.

I think that Kolb viewed organic radicals rather like a pile of marbles held together by glue or perhaps a bunch of grapes all bound to a central stem.

Photograph by Agne27 at the Chaumette Winery, attributed to Don Kasak, source Wikimedia Commons, Creative Commons Attribution 2.0 Generic license.

It should be noted that the stability of the radicals was not inconsistent with contemporary views of ionic behavior since at that time, even inorganic salts were not believed to be dissociated in water.[17] The ability of polar solvents to separate ions was not understood.

---

[17] It was known that solutions of salt conducted electricity, but Michael Faraday had postulated that separation of ions was only a direct result of the electric current. It was not until Svante

Although, in retrospect, Kolbe's view of organic radicals and J.J Thomson's model of the atom now both appear obviously wrong; in both cases, Kolbe and Thompson were interpreting the available data as far as it could go without invoking some sort of mysterious forces.

In my view, the personality of organic chemists (past, present and future) has a large element of *intuition* that is expressed in physical models. At least, organic chemists do not seem to feel limited by existing theories when new data presents itself. As a result, organic chemists in the late 1800s were drawing three-dimensional pictures of molecules and talking about bonds, strain energy, steric effects, optical isomers, resonance, and many other "modern" ideas well before the existence of atoms was completely accepted by physicists.

Kolbe is generally painted as a curmudgeon who at best tried to spoil the party at every turn, and at worst wielded his power to punish individuals who dared to accept the structural theory of chemistry as it was

---

August Arrhenius (1859–1927) wrote his thesis in 1883, that the idea that salts were generally dissociated in solution gained traction.

becoming popular. While the record will show that Kolbe was not a gracious and open-minded colleague; he was right to consistently point out that the structural theory was a complex edifice being built on an island of sand. I would compare his role somewhat as H.C. Brown in the 1960s on the topic of "non-classical carbocations."[18]

Until his death in 1884, Kolbe aggressively opposed what we now know as "structural" organic chemistry every time it arose.

3. Type Theory

Concurrent with radical theory, various chemists were realizing that organic compounds could be viewed as extensions of existing ideas about inorganic molecules. For example, water, alcohols and ethers could all be lumped into compounds of the type R-O-R. This became the guiding principle behind making series of compounds of various types based on inorganic

---

[18] He might also be compared to Peter Duesberg on the relationship of HIV to AIDS in the 1990s.

analogues (water, ammonia, carbon dioxide, alkali metal halides, etc.) The type theory was also used to reconcile the misunderstandings of atomic weights (principally of hydrogen). By assigning a compound to a certain type, a chemist could align its anomalous "gas volume" (i.e., vapor density) with the observations on other compounds of the same type. Type theory evolved into the concept of *functional groups* when the atomic mass system was corrected.

II. Empirical Organic Chemistry (1830-1870)

The absence of theories and instrumentation did not stop innovative chemists in the early 1800s from isolating compounds and discovering chemical reactions in much the same way that alchemists had for a thousand years. The main advances in what I call the "empirical period" was that most mysticism had been removed, results were documented and quantitative chemical analysis was applied. The Empirical Period ended when structural ideas took root in the minds of most organic chemists. This coincided with resolution of several other problems especially lifting the ambiguity about the atomic weight of hydrogen. When organic chemists embraced the structural hypothesis, they were endorsing the atomic theory and the molecular theory of matter and certain thermodynamic concepts that were not yet proven or even accepted as theories by main stream physicists.

1. Justus von Liebig (1803-1873)

I have already mentioned Justus von Liebig (1803-1873) who was aware of the concept of vital force, but apparently had no particular reverence for it. Liebig was introduced to the idea of chemical experimentation at an early age by his father who invented and sold paints and varnishes through his hardware business. In 1816 Liebig experienced the "year without a summer" and the accompanying famine in Germany, which peaked his later interest in agricultural chemistry. His major break came in 1822 when he obtained funds to move from Germany to Paris and study under Gay-Lussac. Here he also made important and supportive older friends (e.g., von Humboldt and Cuvier). In Paris his status was essentially that of a post-doctorate, but he received a doctorate without dissertation from his previous work in Germany in 1823. In 1824, Liebig returned to Germany to become a junior professor at the University of Giessen with minimal funding. However, he was an effective lecturer and received a progressively larger share of the students lecture fees. His career path was opened by the death of two of his older colleagues in 1825.

When the university rejected Liebig's idea for a program to teach industrial chemistry, he was able to found a private institute (while keeping his professorship) and accept students seeking practical education, who would not normally have attended a university. He acquired a small building near the campus and moved his family into the upper floor retaining room for about 10 students to work on the main floor. He operated this establishment for about 10 years (1825-1835). [19] Liebig was well known for his use of analytical methods and formally issued students platinum cups for gravimetric analysis signifying their acceptance to the laboratory. Starting in 1833, Liebig was allowed to implement similar laboratories within the University. One of the innovations he introduced by 1839 was the fume hood.

Some of the personal contributions that Liebig made to organic chemistry include: Synthesis of chloral (1832), identification of the ethyl radical (1834), and oxidation

---

[19] There is an interesting account of the laboratory by one of the students Liebig lured from the university in the period 1834-35. H.G. Good. On the early history of Liebig's laboratory. *J. Chem. Edu.* 557-562 (1936).

of primary alcohols to aldehydes (1835). Obviously, he disagreed with Berzelius in stating:

> *"The production of all organic substances no longer belongs just to living organisms. It must be seen as not only probable, but as certain, that we shall be able to produce them in our laboratories. Sugar, salicin, and morphine will be artificially produced. Of course, we do not yet know how to do this, because we do not yet know the precursors from which these compounds arise. But we shall come to know them."*

In the 1840s, Liebig was particularly interested in application of chemical knowledge to practical problems such as agriculture. He popularized the idea of limiting nutrients (obviously an extension of the idea of limiting reagents in chemistry), but he had a misunderstanding about nitrogen. He assumed that nitrogen in the atmosphere was certainly sufficient for plant growth, but he did not realize that in order for it to be used by plants, it needed to be "fixed" (e.g., converted to nitrate or amine derivatives). This error was eventually corrected in later editions of his books on the subject.

In 1842, Liebig visited England and through his popular lectures, the British realized that they had been surpassed by Germany and France in important areas of technology. Several important British scientists lobbied for establishment of a school for organic and agricultural chemistry. Thus, the Royal College of Chemistry was established in London[20] and Liebig was asked for recommendations to head up the school. In 1845, his student August Wilhelm von Hofmann, of Giessen became head of the school.

In 1852, Liebig was still a relatively young man, but was more interested in chemical education than original research. He, thus, accepted a posh position offered by Maximilian II of Bavaria and moved to Munich. His family went from living over a chemical laboratory to living in a very comfortable home. For the next 21 years (until his death in 1873), he mainly gave lectures and demonstrations, while many of his roughly 700 students went on to be founders of various branches of chemistry.

2. Chemical Analysis

There were few tools available to chemists in the early 1800s, but through the work of Lavoisier, Dalton et al.

---

[20] It operated 1845-1872.

the idea that chemical substances had characteristic combining ratios was taking hold. Of course, the idea of isomerism was poorly understood, in part, because the idea of chemical structures defined by covalent bonds was unknown. Atoms and molecules[21] were debated and heat was regarded by most chemists as a physical substance. Nonetheless, methods of chemical analysis were being derived, (following Lavoisier) primarily gravimetric analysis. In 1810, Gay-Lussac had used potassium chlorate oxidation of organic compounds to measure CO_2 and O_2. Liebig was particularly effective in developing a method for determining the carbon, hydrogen and oxygen content of organic compounds (very similar to that used today). He combusted a weighed portion of the compound of interest and passed the gases over calcium chloride (where water was absorbed) and then over a solution of potassium hydroxide (where carbon dioxide was absorbed). After carefully measuring the increase in weights, oxygen was determined by difference. His technician (Carl Ettling) facilitated these steps by

---

[21] Recall that many diatomic elements were assumed to be monatomic.

fabricating a series of glass bulbs called a *kaliapparat* (i.e., potassium apparatus) in 1831.

Kaliapparat

fotografiert von Terabyte im Liebig-Museum, Gießen, GNU Free Documentation License

Dumas (1833) used an approach similar to Liebig's to estimate the amount of nitrogen in organic compounds (as N_2). The problem, of course, is to remove the excess oxygen from the nitrogen. This could be done by combustion of the gas passing from the H_2O and CO_2 absorbers with excess carbon and again separating the CO_2 and H_2O along with the excess carbon compound.

3. The Dumas Method of Relative Molecular Weights

Once Gay-Lussac established that the density of a gas was associated with its molecular weight, it was not a big step for Jean Baptiste Dumas (1800–1884) to put the principle to work to determine relative molecular weights of volatile compounds. Typically, in the method published in 1826, an excess of a volatile chemical is placed into a glass flask of known volume and weight. The volatile chemical is evaporated at a known temperature and pressure so that it displaces all the air in the flask and is then allowed to condense in the flask. The weight of the condensate is compared to the weight of other (known) materials under the same conditions of pressure and temperature. And the ratio of masses is the ratio of molecular weights.

Dumas Bulb (Fischer Scientific)

4. Products from Plants

For millennia "organic materials" had been used by civilizations without much concern for the theoretical basis of their composition. As chemistry established itself as a science in the early 1800s, those materials were coming under more rigorous analysis. In the following sections, some to the economically important materials that received that attention of chemists are described.

Materials obtained from plants that provided raw materials for chemical investigations included fibers, oils, and aqueous extracts.

Sugars

Green plants make a variety of carbohydrate molecules essentially combining carbon dioxide from the air with water from the soil and expelling oxygen as a byproduct. Of course, the chemistry of photosynthesis was unknown, but the art of fermentation had been developing for millennia by the early 1800s. Naturally, the economics of wine, beer and distilled spirits had received the attention of alchemists and chemists for generations. Following work by Antoine Lavoisier

(1743–1794), in 1810 Joseph Louis Gay-Lussac (1778–1850) was able to write a balanced equation for fermentation (modern formula):

$$C_6H_{12}O_6 \rightarrow 2\ CO_2 + 2\ C_2H_6O$$

At the time, yeasts were considered a non-living catalyst. This position was an issue of concern when the theory of vitalism was proposed, and it was not until 1838 that yeasts were recognized as living organisms. Naturally, Berzelius was pleased that a conversion of organic molecules was being facilitated by living organism, but it was not until after his death that Louis Pasteur (1822-1895) clarified that fermentation was a result of the metabolism of yeast (1850s). The study of sugars lead to two very notable discoveries before 1870:

Anselme Payen (1795–1878) had been studying the fermentation of sugars and in 1833 he extracted the first enzyme from a fermentation broth that breaks down complex (insoluble) starches in barley to soluble sugars. The enzyme was named "diastase" (i.e., separation, because it separated the husk from the starch). Diastase now refers to a family of a hydrolase enzymes (amylase).

Ludwig Wilhelmy (1812–1864) made use of the fact that sugars are optically active in 1850 to follow the acid catalyzed conversion of sucrose solution into fructose and glucose. He developed a rate equation and was able to integrate it. Unfortunately, his contribution was not well understood or appreciated at the time.

Fermentation: Alcohols and Organic Acids

Naturally, the production of wine and beer were fundamental to society from ancient times. And the economic importance lead to significant study to the products of fermentation. Théophile-Jules Pelouze (1807–1867) is mainly known for his students (whom we will meet shortly) rather than his personal contributions to chemistry. He worked on many projects (e.g., atomic weights, glass, military explosives) with well-known scientists. But he is important here for isolating organic acids typically obtained from fermentation of sugars in air. For example, malic acid is the tart component of apples and grapes. It is fermented to lactic acid, which is less-bitter, but some bacteria can carry the process all the way to oxalic acid, which is an even stronger acid and is associated with gout and kidney stones.

L-Malic Acid   L-Lactic acid   Oxalic Acid

Drawings by NEUROtiker and Rjelves,
source Wikimedia Commons

Optical Activity

Soon after the separation of light into polarized rays was discovered by Étienne Louis Malus (1775-1812) in 1808, several experimenters looked for effects of various materials on the resulting rays. In 1815, Jean Baptiste Biot (1774-1862) discovered that turpentine rotates the plane of polarization, but there was no molecular theory that account for why this happened. In 1822, John Frederick William Herschel (1792–1871) discovered that quartz crystals that are mirror images of one another rotated the plane of polarized light in opposite directions. In 1845, Michael Faraday (1791-1867) discovered that the plane of polarized light was rotated by magnetic fields.

As mentioned above, Ludwig Wilhelmy (1812–1864) made use of the fact that sugars are optically active in 1850 to follow the acid catalyzed conversion of sucrose solution into fructose and glucose.

We tend to forget that Louis Pasture (1822 –1895) was a chemist who began his career studying fermentation. In 1848, he discovered that tartaric acid (isolated from dead yeast deposited from fermentation of wine) rotated the plane of polarized light. But tartaric acid made by chemical means did not. When he crystallized the synthetic tartaric acid, he realized that crystals with mirror-image relationships were formed. He then managed to sort the left-handed and right-handed crystals mechanically by hand and found that they rotated the polarized light in opposite directions. In addition, one of the crystals gave a rotation the same as tartaric acid from fermentation. The ability to rotate polarized light was not lost when the crystals were dissolved in water. The fact that the optical activity was not dependent on crystal structure, suggested that it was inherent to the molecules of tartaric acid.

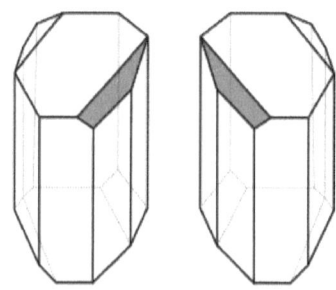

Tartaric acid crystals

Drawing by Brighterorange, source Wikimedia Commons

Dyes and Pigments

Over thousands of years, many different dyes and pigments were obtained by extracting plants.[22] Here I will only mention three that had become important by about 1000 CE in Western Europe: woad, madder and weld.

Weld (*Reseda luteola*) is a European plant introduced into North America as a weed. It produces a flavonoid (luteolin) that can be extracted and used as a yellow dye. When mixed with blue Woad, it provides the color Lincoln Green.

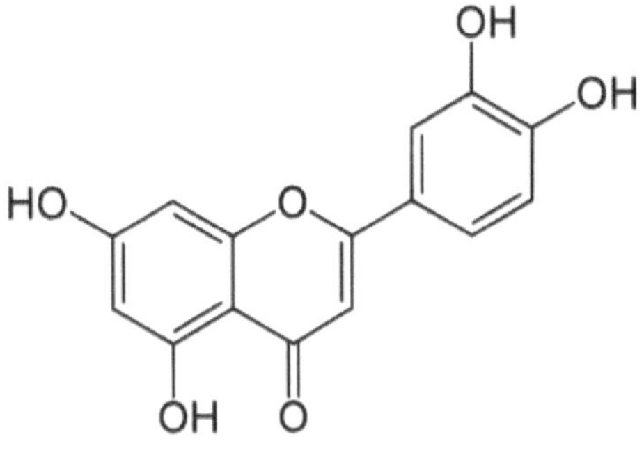

Luteolin

Drawing by Yikrazuul, source Wikimedia Commons

---

[22] M. Sequin-Frey. The chemistry of plant and animal dyes. *J. Chem. Ed.* 58(4) April 1981 pp. 301-305.

Madder is obtained from the roots of *Rubia tinctoria*. Typically, the roots are dug and (after through washing) chopped into very fine pieces, which are extracted with sodium bicarbonate solution at elevated temperature. After filtering to remove the fibrous materials, alum is added and the fabric is dipped in the dye for hours at elevated temperature. The alum helps fix the dye to the cloth. There are two red-colored derivatives of anthraquinone in the extracts:

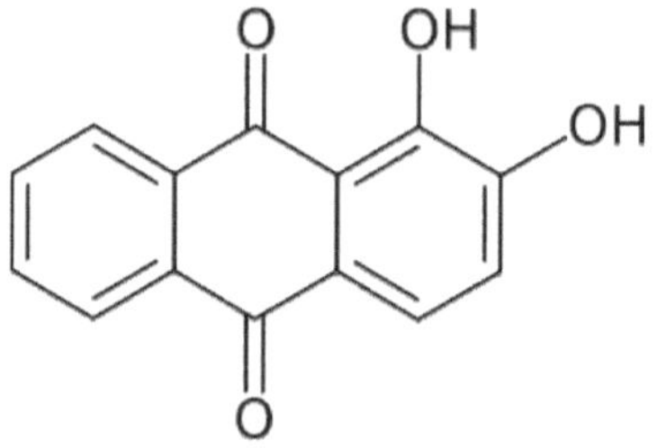

Alizarin (1,2-dihydroxyanthraquinone)

Drawing by Arrowsmaster, source Wikimedia Commons

And

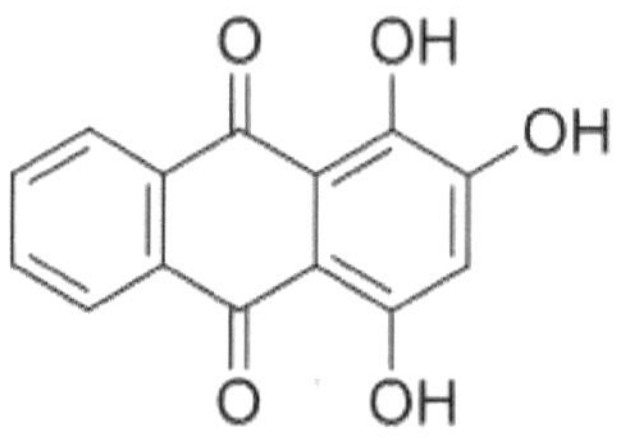

Purpuring (1,2,4-Trihydroxyanthraquinone)

Drawing by Emeldir, source Wikimedia Commons

Woad is a plant (*Isatis tinctoria*) native to Europe that produces low concentrations of a blue dye that is found at higher concentrations in indigo plants, which were originally cultivated in India and in the 1800s in South Carolina. The compound is derived from a glucoside of a derivative of tryptophan. Fermentation of the leaves removed the glucoside and releases indoxyl (1H-Indol-3-ol), which is readily oxidized in the presence of alkali (lye) to indigo.

Indigo

Drawing by Yikrazuul, source Wikimedia Commons

Interestingly, the blue dye is not very soluble in water and is typically reduced to "indigo white" which is placed in the dye bath. After the fabric is soaked in the colorless solution, it is removed to the air where the blue color develops upon air oxidation.

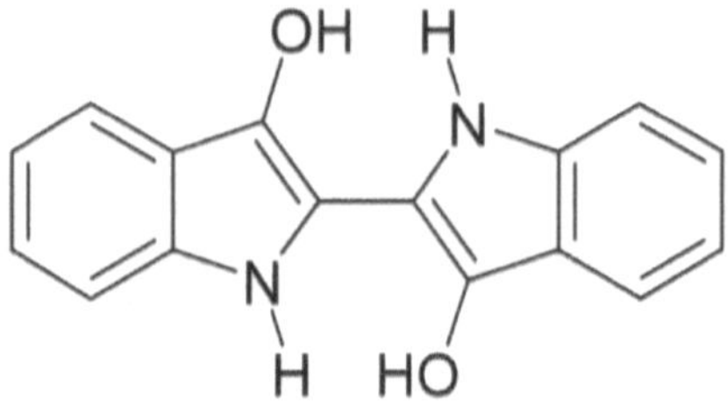

"Indigo White"

Drawing by Klaus Hoffmeier, source Wikimedia Commons

In 1840, it was found that indigo can be oxidized to yield isatin or 1H-indole-2,3-dione by Otto Erdmann (1804–1869) and Auguste Laurent (1808–1853).

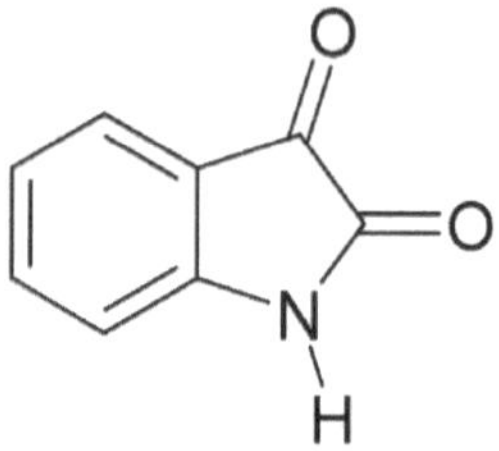

Isatin

Drawing by Klaus Hoffmeier, source Wikimedia Commons

Destructive distillation of indigo by Otto Unverdorben (1806-1873) in 1826 produced aniline, but its composition was not understood until 1843.

Plant Fibers

Plant fibers include primarily lignin and cellulose. Lignin was defined operationally in 1813 by the botanist Augustin Pyramus de Candolle (1778-1841).

He managed to disassemble woody materials into several components and the fibrous component that was insoluble in water and alcohol but soluble in alkali he defined as "lignin" (after the Latin word for wood). Today, we are still trying to clarify the structure of lignin and find a way to utilize it. It appears to be a random polymer containing many phenolic units that forms around plant cells that secrete cinnamyl alcohols (p-coumaryl, coniferyl, and sinapyl alcohols) and other phenolics.[23] Because of its recalcitrance to bio-degradation, lignin and its fragments end up being a major component of soil organic matter (humus).

Cellulose, on the other hand, is systematically biodegradable. It is a regular linear polymer of glucose residues that forms long straight fibrils held together in bundles by hydrogen bonds. The combination of lignin and cellulose make wood a natural thermoplastic. By heating freshly cut (green) wood to about 100ºC, the wood can be made flexible and this technique has been used for centuries to shape the curved hulls of wooden ships. During his study of fermentation, Payen characterized cellulose as a carbohydrate in 1838.

---

[23] R. Hatfield and W. Vermerris. Lignin formation in plants. the dilemma of linkage specificity. *Plant Physiology* 126(4):1351-1357 (August 2001).

Even before cellulose was characterized, it was used as a chemical substrate. In 1832, Henri Braconnot (1780-1855) discovered that it could be nitrated. This initial work only involved low levels of nitration and the product ("xyloidine,") could be dissolved in organic solvent and used as a lacquer (e.g., when the solvent evaporated the cellulose was deposited as a film). Higher levels of nitration in concentrated nitric acid produced flammable paper. By nitrating in a mixture of nitric and sulfuric acids in 1846, Christian Schonbein (1799 – 1868) was able to prepare nitrocellulose (guncotton). Unfortunately, he did not discover how to wash the residual acid from the product and this made his product unstable. Numerous experiments ended in explosions before Frederick Abel (1827–1902) was able to produce stabilized nitrocellulose for smokeless powder. We will see further progress along this line resulting in 1870 with the production of celluloid plastic.

Olive Oil

Obtaining sources of light has been one of the fundamental concerns of humans throughout history and pre-history. Of course, fire provided light and the approach taken in historical times was to find ways to maintain and transport small fires for long periods of time. The olive tree is native to the eastern Mediterranean. The original trees produced small, bitter fruit and since stone lamps suitable for burning olive oil date from 10,000 BCE with manufactured pottery lamps dating from 4500 BCE (6500 years ago, pre-dating the Bronze Age) it seems that the original use of olives was as a fuel/light source. This led to the domestication of the olive tree over 6,000 years ago. Clay jars (amphorae) used to store olive oil date back to nearly the same time period (5,500 years ago, 3,500 BCE). Under selection during domestication, larger and less bitter fruit were obtained and the fruit and its oil became a staple food. Its influence on the development of western civilization should not be underestimated.

Olive oil is composed primarily of triglycerides (triesters of glycerin) with oleic (55-83%) and palmitic (7-20%) acid.

Oleic Acid

Drawing by Ben Mills, source Wikimedia Commons

Palmitic Acid

Drawing by Mrgreen71, source Wikimedia Commons

Latex, Natural Rubber and Isoprene

Rubber was first used in Central America long before contact with Europeans. It was first described by Europeans in 1525 and it was not subjected to serious study until 1735 by Charles de la Condamine. Priestley discovered that it would erase writing and that is where the English name came from. In 1815, rubberized canvas was produced by dissolving rubber in turpentine (later benzole), soaking canvas with the liquid, and evaporating the solvent. This material was employed in waterproof boots and clothing (e.g., the MacIntosh).

To this point, rubber lacked strength and became brittle at low temperatures. In 1840, Charles Goodyear (1800–1860) discovered that heat and sulfur made rubber more durable, but did not immediately patent it. However, in 1842, Thomas Hancock (1786–1865) of the MacIntosh firm acquired a sample of Goodyear's rubber; discovered how Goodyear had made it and patented "vulcanization." Vulcanized rubber immediately found important markets such as toys and the pneumatic tire (1845), which was widely utilized in the early bicycles (1869).

Throughout this period, the natural source of rubber was very difficult to grow and harvest. The British transplanted trees from Central America to South America, India, Africa and Southeast Asia. But it took long periods to grow the trees and harvesting the latex was literally done drop by drop. Obviously, interest in the composition of rubber and methods of its synthesis grew in Europe.

In the 1800s, the composition of rubber was unknown as was its biosynthetic pathway:

Drawing by Smokefoot, source Wikimedia Commons

In 1860, Charles Greville Williams (1829–1910)) destructively distilled natural rubber and obtained a light oil (b.p. 37-38°C[24]), which has the same composition as natural rubber (88.23% carbon and 11.77% hydrogen) and which he called "isoprene" (C_5H_8). However, the structure of isoprene was not understood until 1882 through the work of William Tilden (1842–1926).[25]

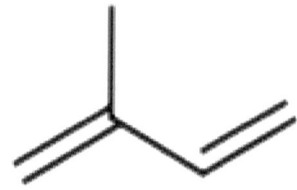

Isoprene (C_5H_8)

Drawing by Fvasconcellos, source Wikimedia Commons

Tar, Pitch and Turpentine

Into the 1800s, national prestige was tied to wooden sailing ships. More than any other nation of the period,

---

[24] Modern value 34°C. See page 171 of F.J. Pond. A review of the pioneer work on the synthesis of rubber. *J. Amer. Chem. Soc.* 36:165-199 (1913).

[25] F.J. Pond. A review of the pioneer work on the synthesis of rubber. *J. Amer. Chem. Soc.* 36:168 (1913).

Britain depended on imports from its empire to maintain its navy. The massive trees needed for masts and spars were in short supply in Britain relative to the continental powers (France and Spain). In the early 1700s, the British received a large amount of their timber from the Baltic States. The North American colonies became the preferred source of "naval stores" in the late 1700s and early 1800s although this was punctuated by the War of Independence (1775-1780) and War of 1812. The French and British both introduced iron ships in 1860-61. In the American Civil War (1861-65), we see the transition from wood to steel and sail to steam. Because of its needs for raw material exports, the southern American States were eager to trade with the British and North Carolina (lacking the Tobacco and Cotton exports of Virginia and South Carolina-Georgia, respectively) became to the center of the trade in tar, pitch and turpentine. Hence, the nickname "Tar Heels" and towns with names like "Tarboro." The long leaf pine forest of the Carolina coastal plain with ready access to shipment from Wilmington and Charleston was the scene of much of this activity.

The ecology of the pine tree is the source of its resins. The trees grow fast and dense reaching for sunlight in thick stands. They have few lower branches. The trees can reach considerable height in as little as 20 years and their main enemy is fire.[26] When fire strikes in the underbrush, pine straw and weeds, the lower part of the tree is burned and the copious sap flows downward to protect the tree from heat. The resin carbonizes to tar, pitch and carbon, which helps insulate the tree allowing it to survive during these fast-moving burns.

The resin is composed of terpenes which are conceptually derived from isoprene (C_5H_8):

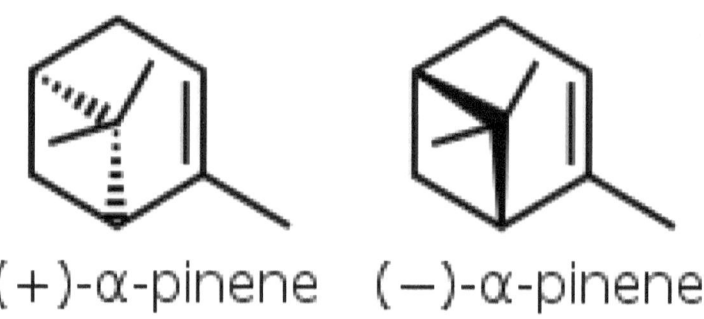

(+)-α-pinene (−)-α-pinene

Drawing by Inductiveload, source Wikimedia Commons

---

[26] Unlike longer-lived trees (e.g., cedars and oaks) they produce no particular anti-insect or anti-fungal natural products.

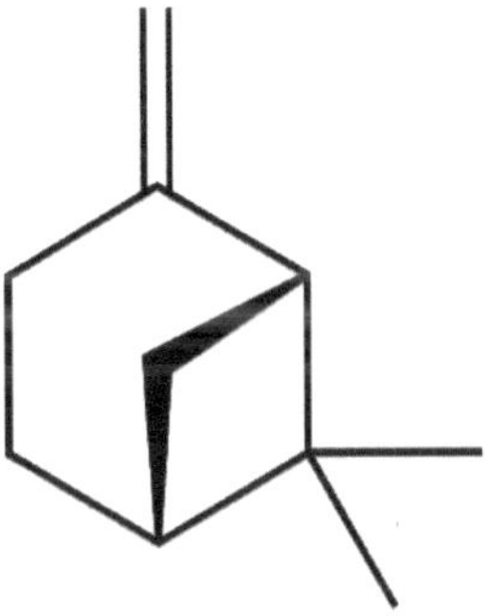

β-pinene

Drawing by Ben Mills, source Wikimedia Commons

The biosynthetic pathway is as follows:

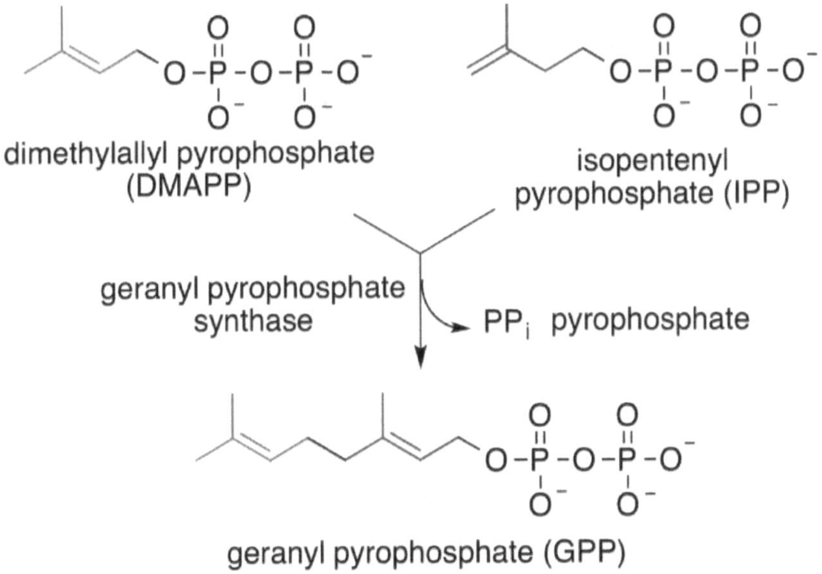

dimethylallyl pyrophosphate
(DMAPP)

isopentenyl
pyrophosphate (IPP)

geranyl pyrophosphate
synthase

PP_i pyrophosphate

geranyl pyrophosphate (GPP)

Drawing by Andrew Murkin, source Wikimedia Commons

Followed by:

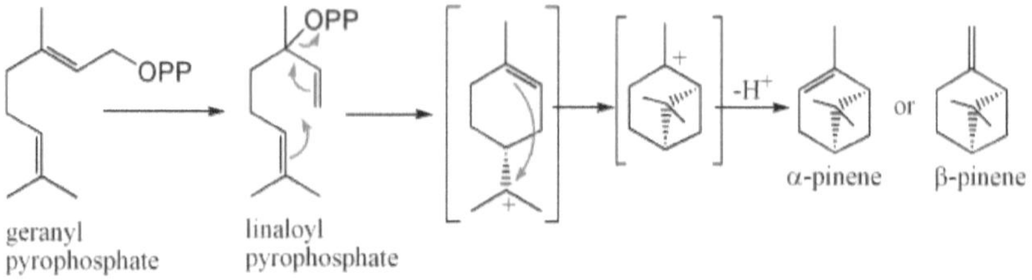
OPP
OPP
-H+
or
α-pinene
β-pinene
geranyl pyrophosphate
linaloyl pyrophosphate

Drawing by Walkerma, source Wikimedia Commons

Related compounds (e.g., delta-3 carene, sabinene, limonene and terpinolene) are also present. It is interesting to examine the mechanism of formation of limonene to see how these compounds are related:

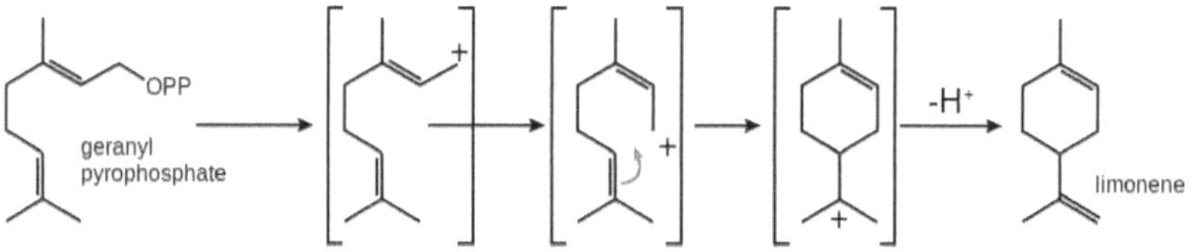
OPP
geranyl pyrophosphate
-H+
limonene

Drawing by Fred the Oyster, source Wikimedia Commons

If the pine resin is collected much like latex, it can be distilled to produce a liquid (optically active) made up primarily of the pinenes (45-75% alpha; 5-30% beta) and the residual pitch is left in the still.

As noted above, turpentine was a well-known material in the 1800s and it was readily isolated in quantity from the sap of pine trees. In 1815, Jean-Baptiste Biot discovered that turpentine caused optical rotation of polarized light, but without a structural understanding, no one knew what that meant. In 1818, the French botanist Jacques Labillardière (1755–1834) distilled turpentine looking for its constituent compounds. Fractional distillation of turpentine separates it into two major components: α-pinene (b.p. 155°C) and β-pinene (b.p. 166°C). Chemical analysis (C/H/O) gives the empirical formula (C_5H_8), and molecular weight by colligative property measurements and comparison of boiling points indicates that the molecular formula is $C_{10}H_{16}$. Other oils with the same empirical formula were discovered over the next decades. Dumas had been working with camphor and determined its formula to be $C_{10}H_{16}O$ in 1833. There is some disagreement in the secondary literature whether Dumas or Kekulé (1866) first used the term "terpene" to distinguish these C_5H_8 compounds from camphor compounds. The structures of the terpenes (pinenes) was deduced in the 1880s after the structure of carbon compounds was understood.[27]

---

[27] G. Singh. Chemistry of Terpenoids and Carotenoids, Volume 3

Quinine and Artemisinin

The bark of the cinchona tree (known locally in the Andes Mountains as the "quina") was known to reduce fevers and shivering from cold long before Europeans arrived. The history is murky, but it is alleged that in the 1500s the wife of the Spanish Viceroy in Peru (a.k.a., the Chinchon Countess) contracted malaria and was cured of the disease by a native doctor with the bark. A Jesuit priest Bernabé Cobo (1582–1657) is credited with introducing the plant and bark to Europe in 1632. It soon was found effective against malaria (which was common in southern Europe especially around Rome) and its value was established. Thus, for many years, Peru attempted to prevent export of the trees or its seeds, thus maintaining a monopoly on the drug. Eventually, European merchants and governments managed to steal seeds and start plantations in various tropical regions. The most successful plantations were in Southeast Asia.

Because of its unique anti-malarial activity, quinine became important to the exploration and exploitation of Africa and other tropical regions by Europeans in the

1700s and 1800s. Initially, the dried bark was chewed or used to prepare a tea, which was administered to the patient.

Quinine

Drawing by CYL, source Wikimedia Commons

Quinine was isolated in crystalline form in 1820 by Pierre-Joseph Pelletier (1788-1842) and Joseph-Bienaime Caventou (1795-1877). In 1834, Carl Warburg (c. 1805–1892) introduced an alcoholic extract of a native plant in Guyana (then British Guiana), which proved to have quinine as a major component. Its economic importance encouraged chemists to attempt its synthesis. Indeed, the young William Henry Perkin (1838–1907) was looking for a synthesis of quinine when he accidentally made mauveine (1856).

While Europe focused on quinine, in China the focus was on the botanical isolated from *Artemisia annua* (sweet wormwood, which has now been spread worldwide). This compound is relatively fragile molecule and needs to be extracted from the plant using cold water or alcohol (not boiling water). As resistance to quinoline-based drugs has increased (post-1970) artemisinin has become more important.[28]

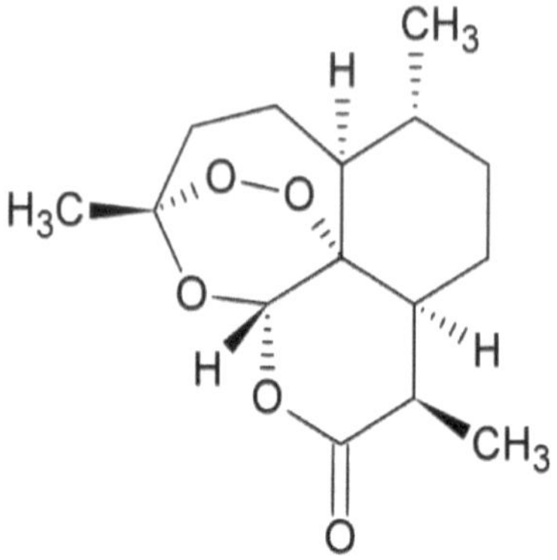

Artemisinin

Drawing by Lukáš Mižoch, source Wikimedia Commons

Nicotine and Niacin

Tobacco had been introduced into Europe in the 1500s. Naturally, chemists became interested in the active ingredient. Nicotine was isolated from tobacco in 1828,

---

[28] Martino E. et al. Artemisinin and its derivatives; ancient tradition inspiring the latest therapeutic approaches against malaria. *Future Med Chem.* 2019 Jul 12.

but it was not chemically characterized until 1843 by Louis Melsens (1814–1886).

In 1891, the empirical formula was clarified by Adolf Pinner and Richard Wolffenstein (1864–1926).[29] By 1904, chemists were attempting synthesis of nicotine via substituents on the pyridine ring.[30]

Nicotine
Drawing by Harbin, source Wikimedia Commons

In 1873, Hugo Weidel (1849–1899) isolated niacin by oxidation of nicotine with nitric acid. Niacin turned out to be an important vitamin (vitamin B3 or nicotinic acid). Absence of niacin in the diet causes the disease known as pellagra.

---

[29] A. Pinner and R. Wolffenstein. Concerning nicotine. *Berichte der deutschen chemischen Gesellschaft.* 24(1):1373–1377 (1891).

[30] A. Pictet and A. Rotschy. Synthesis of nicotine. Berichte der deutschen chemischen Gesellschaft. 37(2):1225–1235 (1904).

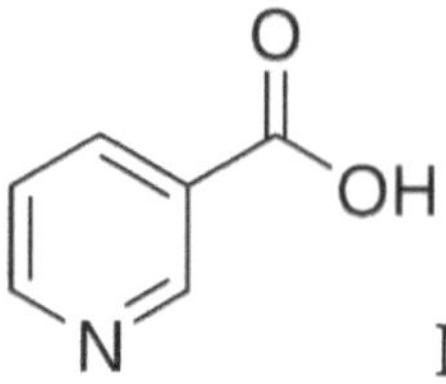

Niacin

Drawing by Harbinary, source Wikimedia Commons

Salicin

Extraction of the bark of willow and poplar trees to obtain a fever- and pain-reducing drink goes back thousands of years and was mentioned by Hippocrates (460-377 BCE). Native Americans were doing it at least a thousand years ago and The Reverend Edward Stone has been credited with mentioning it in England in 1763. In 1826, Henri Leroux isolated the active ingredient from willow bark and in 1828 the material was named salicin after the Latin name of the willow (*Salix alba*) by Johann Büchner[31] (1783–1852).[32] Salicin is

---

[31] Johann Buchner had a son named Ludwig Büchner (1824–1899) and he had a nephew Ernst Büchner (1850–1924) who patented the Büchner funnel in 1888.

[32] J. Miner and A. Hoffhines. The discovery of aspirin's antithrombotic effects. *Texas Heart Institute Journal* 34(2):179-186 (2007).

a β-glucoside, which was first hydrolyzed by Raffaele Piria (1814–1865) in 1838.

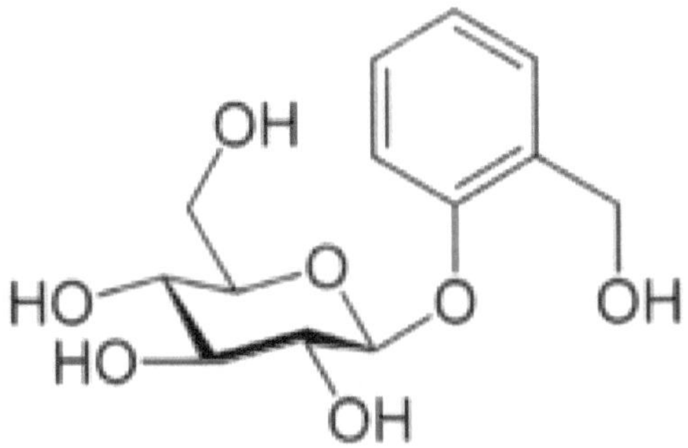

Salicin

Drawing by Ben Mills, source Wikimedia Commons

The side-chain of the phenol is oxidized under mild conditions and *in vivo* to yield salicylic acid.

Charles Gerhardt (1816-1856)[33] was making anhydrides of many acids by reaction of their salts with acetyl chloride (CH$_3$COCl) and applied his technique to a salt of salicylic acid. He assumed the product he isolated in 1853 was an anhydride and did no follow up studies. In 1859, von Gilm did essentially the same thing with

---

[33] He must have been a very annoying person as he had bitter disputes with his father, mentors (Kolbe, Liebig and Dumas) and colleagues. He ultimately (accidentally) poisoned himself with his own chemicals before he was 40 years old.

the acid itself. It was not until 1869, that the products made by Gerhardt and von Gilm were shown to be the same thing and that they were the acetyl esters of the phenol, not anhydrides.

Furfural and Catalytic Hydrogenation

Johann Döbereiner (1780–1849) was born into a poor family and obtained his introduction to chemistry as an apprentice to an apothecary. In the process of isolating formic acid by distilling ants, he apparently isolated a small quantity of furfural.[34] In 1840,

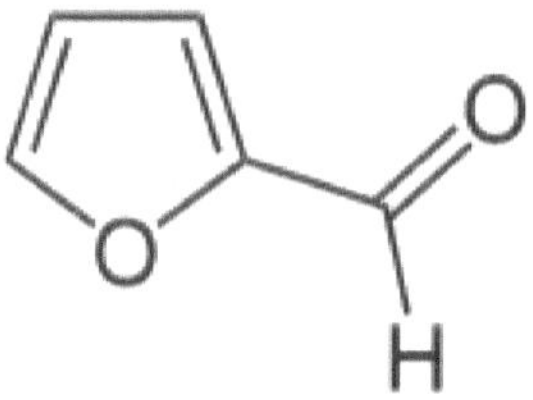

Furfural

Drawing by NEUROtiker, source Wikimedia Commons

John Stenhouse (1809–1880) found that it could be distilled from a variety of grains, and in 1901, Carl Harries (1866–1923) deduced its structure. Of course,

---

[34] J.W. Döbereiner. *Ueber die medicinische und chemische Anwendung und die vortheilhafte Darstellung der Ameisensäure.* *Berichte der deutschen chemischen Gesellschaft* 3(2):141–146 (1832).

the Quaker Oats Company turned it into an industrial commodity in 1922.

But, Döbereiner's most important contribution was use of platinum metal as a catalyst. He used it to ignite a stream of hydrogen produced by reaction of zinc and acid in a household appliance used to ignite fires.

5. Products from Animals

Soap and Candles

While olive oil was preferred for cooking and lighting, it was found that animal fats were also flammable. In particular, tallow (animal fat) was found to burn. In contrast to olive oil with oleic acid, animal fat is principally composed of steric acid triglycerides.

Steric Acid

Drawing by Slashme, source Wikimedia Commons

Typical tallow constituent

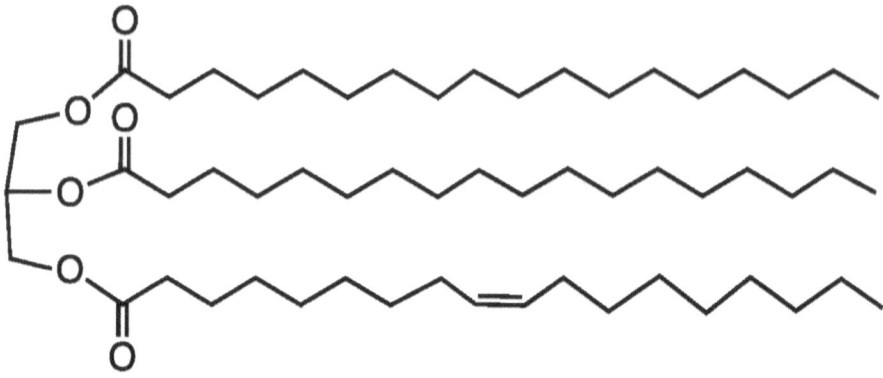

Drawing by Smokefoot, source Wikimedia Commons

Candles were invented by placing a strip of plant fiber into oil and using it as a wick to control the rate of combustion. In Roman times solid candles were developed using bee's wax, but these were far too expensive for the ordinary people. In the 1600s and 1700s, whale oil wax was used for candles, but these were also expensive.

Meanwhile, ancient peoples had learned that oils (e.g., olive oil) and salts were effective in cleaning the skin and hair. The Phoenicians were apparently the first to mix wood ash (potash) and tallow and actually hydrolyze the glycidyl esters (i.e., saponification about 600 BCE). By Roman times, saponification was known and soap (e.g., salts of steric acid) was fairly well known within the empire. The alchemists found that mixing soap with (sulfuric) acid produced the free steric acid.

In 1818, Henri Braconnot (1780-1855) obtained a patent for candles made from stearin (glyceryl tristearate), which melted between 50 and 70ᵒC. In 1825, Michel Eugène Chevreul (1786–1889) and Gay Lussac patented a process for making candles from steric acid.

Creatine

Chevreul (1786–1889) was very active in extracting products from animals[35] and isolated creatine in 1832. Creatine is a source of quick energy (synthesis of ATP) and is found in high concentrations (~1%) in the muscles of warm-blooded animals.

Drawing by Edgar181, source Wikimedia Commons

---

[35] *Recherches sur les corps gras d'origine animale* (1823)

Glycerin

Glycerin was originally isolated as a byproduct of soap making. During the 1840s, there was interest in replacing (or supplementing) black powder with other explosives. Théophile-Jules Pelouze at the University of Paris was involved in the study of many things including nitrocellulose. In 1838, Pelouze had dipped cotton balls in concentrated nitric acid and found that the product would explode. In 1846, Christian Friedrich Schonbein (1799-1868) accidentally discovered that a higher level of nitration of cotton (cellulose) could be achieved by using a mixture of sulfuric and nitric acid.

The success of guncotton (nitrocellulose) spurred more investigation among Pelouze's former students Ascanio Sobrero (1812–1888) and Alfred Bernhard Nobel (1833-1896). In particular, Sobrero (then a professor at the University of Turin) produced nitroglycerin (a tri-nitric acid ester of glycerin) in 1847, but he soon found it so unstable that he abandoned work on it.

The Nobel family (who had moved from Stockholm to St. Petersburg) had manufactured munitions for the Russians during the Crimean War (1853-1856), but after

the war the family's business neared bankruptcy.
Nobel was a student of Pelouze in the early 1850s
where he learned explosive technology. Then he spent
time in the US during the run-up to the American Civil
War (1861-65). He was probably looking for business
opportunities for industrial explosives (e.g., tunnels,
canals) in the expanding American economy. In 1859,
Alfred Nobel and his parents returned to Sweden,
leaving the plant in Russia to Nobel's older brother. In
Sweden, Nobel knew that nitroglycerin was a powerful
explosive and very easy to make, but its instability
limited its utility. Nonetheless, the family proceeded to
make and ship quantities of nitroglycerine as an
industrial explosive. Unfortunately, Alfred Nobel's
younger brother was killed in an explosion of a
production plant in 1864.

Nobel apparently came to the idea that if he could find
a way to detonate low concentrations of nitroglycerin
(which were relatively stable), he could make a safe
explosive. To this end, he focused on inventing a
detonator (i.e., blasting cap), which he invented in 1863.
Nobel perfected his detonator in 1865. For two more
years, Nobel looked for a way to stabilize nitroglycerin
so that it would be convenient to handle. He
discovered that when it soaked into diatomaceous

earth (which was used to pack around bottles of the liquid for shipment) it was very stable to heat and shock. Thus, he invented dynamite in 1867. This invention was patented in numerous countries and Nobel soon became a wealthy man.

Additional developments in explosives followed into the 1880s. In his old age, he was influenced by pacifists and realized that he might go down in history as a villain for his invention that was closely associated with mass death and destruction in the much more lethal wars being fought. Moreover, collateral deaths of civilians and destruction of their property incidental to battles was eroding the romance of war.[36] Thus, (lacking heirs) Nobel directed that after his death, his fortune should be used to establish prizes for peace and scientific achievements, i.e., the Nobel Prizes.

---

[36] The burning of civilian homes and cities had been largely abandoned in Europe after the feudal period. Even in the Napoleonic wars, only Moscow was burned and it was burned by the defending Russians to deny the French winter quarters. Open war on civilians was reopened by the burning of Atlanta (GA), many plantations in South Carolina, Columbia (SC), and Richmond (VA) during the American war (1861-65) although the Union propaganda claims in each case that the Confederates burned their own homes. W.T. Sherman: "War is hell!"

6. Products from Petroleum and Coal

Petroleum

Petroleum seeps were known in various areas including in some coal mines. Samples were periodically distilled and petroleum wax was discovered in 1830.

Abraham Gesner continued his research on fuels and wrote a number of scientific studies concerning the industry including an 1861 publication titled, "A Practical Treatise on Coal, Petroleum and Other Distilled Oils," which became a standard reference in the field.

Coke and Coal Tar

Britain and Germany led European countries in the natural resource of coal. The manufacture of iron required coke (essentially pure carbon) to reduce iron oxide ore. Coke was produced by heating coal with minimal air with the loss of hydrogen, oxygen, nitrogen and sulfur as various (thermally stable and volatile) compounds. Production of coal gas for lighting was also widely practiced and the principal residue was known as "coal tar." A number of people experimented with coal tar. For example, in 1825 Faraday had collected *bi-carburet of hydrogen*[37] from a coal oil, which turned out to be benzene. Liquid fuels were desirable; but until 1850 Americans and Western Europeans depended on whale oil while in the Mediterranean olive oil was traditionally used. These oils were becoming more difficult to find and their prices were being driven up beyond the means of most people who relied on wood and coal for heat. Various types of coal-derived liquids (byproducts of iron making) were known to burn, but generally these were smelly and smoky (highly aromatic) materials not suitable for domestic use.

Thus, in 1846, Abraham Gesner (1797–1864) in Canada found a way to optimize the yield of a combustible liquid from coal and coal-like materials (e.g., shale).[38]

---

[37] Remember that at the time, hydrogen was not understood to be diatomic so the empirical formula would today be C_2H_2 which would be "bi-carburet of hydrogen."

[38] Details are sketchy, but the process seems to have used a retort (which limited oxygen) and possibly use of water or liquid oils

Gesner's product was superior because it contained mainly aliphatic (high-hydrogen to carbon ratio) compounds, and, thus, it burned with less smoke and had less odor. He named his product "kerosene," which was patented as a trade name (not as a chemical name), and built a successful business in 1850 (i.e., The North American Gas Light Company). Similar work was undertaken in Britain by James Young (1811–1883) and there were disputes over intellectual property. However, both companies were doomed by the work of Samuel Kier (1813–1874) who discovered a method of making a comparable product (called "carbon oil") from crude petroleum in 1851. Edwin Drake (1819–1880) made petroleum commercially available in 1859 by drilling a well in Titusville, Pennsylvania for the Seneca Oil Company. Drake's main contribution was the idea of driving an iron pipe through the collapsing sand and gravel overburden to bedrock and then drilling through the pipe and bedrock with a steam-powered drill until an oil seam was encountered. From here, petroleum soon became available and coal was displaced as a source of kerosene. We will pick up the

---

to increase the hydrogen content. Most sources just refer to distillation of coal. The process became obsolete with the commercial production of petroleum.

story of petroleum and petroleum chemistry later, but first we should look at the chemistry that was being done with coal tar before 1870.

Benzole (Benzene-Toluene Mixture): Simple distillation of coal tar yields a variety of hydrocarbons, but benzole (a mixture of benzene and toluene) is one of the principal components. Naphthalene is also obtained.[39] These compounds are relatively small and particularly stable. Thus, when complex hydrocarbons of coal are pyrolyzed, it is not surprising that they would be formed (e.g. possibly via levoglucosan ($C_6H_{10}O_5$) a pyrolysis product of starch in the presence of reducing agents like H_2S, or hydrocarbons undergoing free radical H_2 elimination. See Bergius process).[40]

Benzene and Nitrobenzene: Eilhardt Mitscherlich (1794-1863)[41] taught at the University of Berlin (1821-1863).

---

[39] Charles Mansfield (1819–1855) was a student of von Hofmann (1848) and improved the yield of benzole from coal tar.

[40] M.R. Nimlos and R. J. Evans, Levoglucosan Pyrolysis. Fuel Chemistry Division Preprints of the National Renewable Energy Laboratory 47(1):394-396 (2002).

[41] Among other work, he produced diethyl ether from ethanol and realized that (while acid was required to cause the reaction) acid was not consumed in the reaction. After reviewing these studies, Berzelius introduced the term "catalyst" into chemistry.

He isolated pure benzene (*benzin*) while attempting to distill benzoic acid from calcium hydroxide (slaked line). He recognized benzin (e.g., phenyl) as a radical and this created a conflict with Berzelius (see above) who believed that benzoyl was the relevant radical. Among various reactions of benzene, Mitscherlich nitrated benzene with concentrated nitric/sulfuric acid in 1832 and published his work on benzene derivatives in 1834.

Aniline and Benzidine: As noted above, aniline had been isolated as a decomposition product of indigo in 1826, but it was first extracted from coal tar in 1834 by Friedlieb Runge (1795-1867). In 1842, Nikolay Zinin (1812–1880) obtained aniline from reduction of nitrobenzene with sodium sulfide.[42] But, principal credit probably goes to August Wilhelm von Hofmann (1818-1892) who showed that these were all the same material in 1843. Zinn and Hofmann were students of Liebig. In 1845, Zinin reported the synthesis of benzidine from nitrobenzene by reduction with H_2S in ammonia (via, azoxybenzene, azobenzene,

---

[42] On a commercial scale, iron powder and hydrochloric acid form the preferred reducing agent used after 1850.

hydrazobenzene) followed by treatment with sulfuric acid, which facilitates the rearrangement.

Phenol and quinoline: Quinoline was first isolated by Ferdinand Runge (1795-1867) from coal tar in 1834. A variety of quinoline-based drugs have been synthesized (see below).

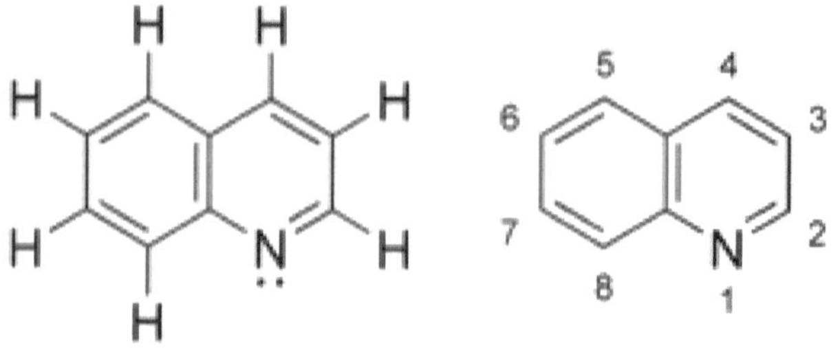

Quinoline

Drawing by cacycle as modified by Walkerma, source Wikimedia Commons

Mauveine (a.k.a., aniline purple): The first aniline dye was accidentally synthesized by William Perkin (1838–1907) a student of von Hofmann in 1856. Perkin was only 18 years old and was trying to make synthetic quinine by oxidizing a mixture of aniline and o,p-toluidines[43] with potassium dichromate. There are actually four separate compounds with methyl groups

---

[43] From the nitration and reduction of benzole, see above.

in different positions resulting from incorporation of toluidine in different places. The reaction leads to a black tar, but when cleaning the glassware with organic solvents, Perkin noticed that some of the extracts were purple and the color adhered to textiles. Perkin patented the product and set up a company that soon was very profitable.

Mauveine A (one of four compounds)

Drawing by Reubot, source Wikimedia Commons

Thiophene: Crude benzene (b.p. 81°C) from coal tar or petroleum was recognized to be mixed with toluene and xylenes, but these could be separated by distillation yielding what was assumed to be pure benzene. In 1883, Victor Meyer (1848-1897) was working with isatin (the oxidation product of indigo), which was known to form a blue dye with crude benzene. But he realized that the blue color was not a reaction with benzene but rather an impurity in benzene. That impurity was isolated and found to be a sulfur compound now called thiophene (b.p. 84°C)

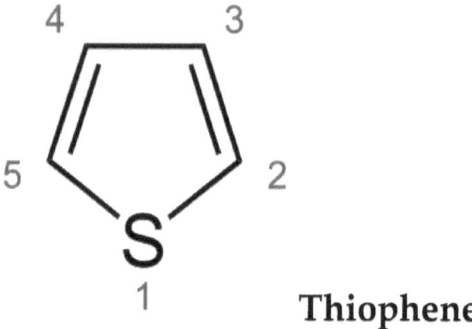

Thiophene

Drawing by Jynto, source Wikimedia Commons

8. Synthetic and Structural Organic Chemistry

Starting in 1854, Marcellin Berthelot (1827-1907) founded the science we now call *synthetic organic chemistry*. Up until this time, chemistry and physics had been thoroughly comingled in the study of the atom and chemistry consisted primarily of isolating new metals (e.g., Davy, Faraday and Berzelius) or naturally occurring organic compounds. Indeed, it was generally believed (based on Berzelius) that "organic" compounds were the unique products of living systems. Berthelot proved that to not be the case as he devised methods of synthesis for various "organic compounds" by inorganic means.

Ironically, an equally argumentative and close-minded chemist came to the forefront. Hermann Kolbe (1818–1884) had helped break down the theory of vital force, and saw the utility of radicals, but he was thoroughly wedded to Berzelius's focus on electrostatics in radicals. Kolbe contributed to the science of organic synthesis (indeed, he coined the term "synthesis") and watched the radical theory expand.   But Kolbe did not grasp and strongly resisted the ideas of the structural theory of (organic) chemistry, which was growing up parallel to organic synthesis. These debates on scientific issues, unfortunately, often became personal attacks. Even luminaries like Justus von Liebig and Jean Baptiste Dumas sharply disagreed periodically.

Organic Amines

Charles Wurtz (1817-1884) was a student of Justus von Liebig (1842) and Jean Baptiste Dumas (1843) and a friend of von Hofmann whom he met in Liebig's lab. In many ways Wurtz's background was similar to Kolbe and he supported the concept of organic radicals; but contrary to Kolbe, Wurtz absorbed and defended the structural theory of organic chemistry that developed in the 1860s. He made personal contributions in several areas.

Wurtz had isolated ethyl isocyanate (Et-NCO, mis-identified as "ammoniacal cyanic ether" implying Et-OCN) and in 1849, he found that it could be hydrolyzed to ethylamine. This was a surprise because Frankland and Kolbe had hydrolyzed ethyl cyanide to propionic acid and ammonia. He published his discovery and received warm congratulations from Kolbe (who had predicted the existence of alkylamines based on radical theory). Ammonia was regarded as the "type" upon which all organic bases were built. In England, von Hofmann realized that this was the explanation for the basicity of aniline and although he was disappointed that he had not grasp the idea first, he responded by publishing the synthesis of secondary and tertiary amines in late 1849. Liebig was very pleased with his former students and wrote to Hofmann (23 April 1849):

> *". . . every new compound is the first member of a new series of homologous compounds, and the thought from which the compounds arose is like a seed corn, which bears its fruits in the minds of others doing similar work."* [44]

By, 1853, Dumas retired and Wurtz took over his chair.

Meanwhile, Adolph Strecker (1822–1871) was also experimenting with cyanohydrins. In 1850, he found[45] he could make alpha-amino acids (basic building blocks of proteins) by this method (now known as the Strecker sysnthesis:

---

[44] Cited from Alan J. Rocke. The Quiet Revolution: Hermann Kolbe and the Science of Organic Chemistry. UNIVERSITY OF CALIFORNIA PRESS (1993) p. 97.

[45] A. Strecker. *Ueber die künstliche Bildung der Milchsäure und einen neuen, dem Glycocoll homologen. Annalen der Chemie und Pharmazie* 75 (1): 27–45 (1850).

Drawing by Krishnavedala, source Wikimedia Commons

Aromatic Diazonium

After a lackluster academic history, Johann Peter Griess (1829–1888)[46] worked briefly for Kolbe in 1855, but he was not pleased with Kolbe's discipline and soon left to work in the Oehler factory fractionating coal tar to recover aniline. Unrelated to his work, benzene vapor caught fire and burned the factory down forcing him to return to Kolbe's laboratory. It is amazing the effect that a dirty job can have on the academic motivation of a young man. This time in Kolbe's lab, his work ethic was much improved and he focused on aniline. He became aware that in 1848, a paper had been published by Raffaele Piria (1814–1865) describing the reaction of aspartic acid with HNO_2 (nitrous acid) to produce malic acid by deamination (probably via fumaric acid). In Hofmann's laboratory at the Royal College of

---

[46] V. Heines. Peter Griess discoverer of diazo compounds. *J. Chem. Educ.* 35:187–191 (1958).

Chemistry, similar reactions had been used to displace amine groups from aniline and aminobenzoic acid by 1853, but after Perkin's discover of Mauveine in 1856, work in von Hofmann's lab had turned to dyes. Griess began a systematic study of the reaction of nitrite ion (NO_2^-) with aniline in acid. By the spring of 1858, Griess published "A Preliminary Notice on the Influence of Nitrous Acid on Aminonitro- and Aminodinitrophenol" in Liebig's *Annalen*. This happened at the same time that Kolbe had recommended Griess to von Hofmann as an employee. Griess had discovered that 4,6-dinitro-2-amino-phenol (which was easier to obtain that aniline) precipitated an explosive azo-compound from alcohol.

Hofmann and Griess came to terms and Griess spend the next three years isolating diazonium compounds (including the yellow compound diazobenzoic acid) and studying their reactions (Published 1860-66). He left Hofmann for a better paying job at Allsopps and Sons Brewery in 1862. In 1864, Griess published a paper describing the coupling of doubly diazotized benzidine with aromatic compounds. But, Griess (unlike Perkin) seems to have been content with the science.

Meanwhile Roberts, Dale and Company of Manchester had found a yellow dye for dying cotton. Another one of Hofmann's students, C. A. Martius, identified this as Griess's diazobenzoic acid in 1866. That same year, another yellow dye also appeared from Simpson, Maule, and Nicholson, which was identified as p-phenylazoaniline by Griess and Martius. It was soon apparent that a general route to colored compounds with good dye characteristics was provided via coupling of diazotized aromatic amines with phenol or anilines. Within a few years, dozens of combinations were produced and the lucrative diazo dye industry was creating great interest in the nature of the benzene/phenol/aniline structure.

The Sandmeyer Reaction

In 1884, Traugott Sandmeyer (1854–1922) working for the company that would come to be named Ciba-Geigy published two papers[47] on the substitution of an

---

[47] Sandmeyer, T. Ueber die Ersetzung der Amidgruppe durch Chlor in den aromatischen Substanzen.
Ber. Dtsch. Chem. Ges. 17, 1633–1635 (1884).

Sandmeyer, T. Ueber die Ersetzung der Amidgruppe durch Chlor, Brom und Cyan in den aromatischen Substanzen.

aromatic amino group for other substituents using Cu(I) salts. Typically, the diazonium is formed at 0°C in the usual way and brought to near neutral pH and warmed to about 60°C with addition of the appropriate copper(I) salt, e.g., CuCl. The reaction mechanism appears to be a series of single electron transfers (SETs): Cu(I)X is oxidized to Cu(II)X$_2$ as it transfers an electron to the diazonium, which loses N$_2$ producing an aryl radical that abstracts X from CuX$_2$ such that Cu(I)X is regenerated (i.e., Cu(I) is catalytic). An aryl diazonium can also produce biaryl compounds (in modest yields) via the Gomberg-Bachmann reaction.[48]

The Kolbe-Schmitt Reaction

Meanwhile Rudolph Schmitt (1830-1898) came to work for Herman Kolbe as a student in 1857 and became friends with Johann Griess. In 1861, he discovered that sulfuric acid reacts with aniline to form N-phenylsulfamic acid, which rearranges (at 180-190°C) to

---

Ber. Dtsch. Chem. Ges. 17, 2650–2653 (1884).

[48] M. Gomberg and E.W. Bachmann. The synthesis of biaryl compounds by means of the diazo reaction. J. Am. Chem. Soc. 42 (10): 2339–2343 (1924). Also see H. Meerwein, E. Buchner and J. van Emster. Prakt. Chem. 152: 237 (1939).

sulfanilic acid (4-aminobenzenesulfonic acid). Apparently, this success led Kolbe and Schmitt to attempt to conduct similar reactions with carbon dioxide and sodium phenolate (i.e., sodium salt of carbolic acid) in 1863. They succeeded only at high temperature (125°C) and pressure (100 atm) and isolated 2-hydroxybenzoic acid (i.e., salicylic acid). At the time, this was purely an exercise in chemical synthesis with no specific economic interest:

Drawing by Benjah-bmm27, source Wikimedia Commons

Alcohols and Ethers

Meanwhile, Alexander Williamson (1824–1904) was perfecting a technique to prepare ethers from alcohols between 1850 and 1852.

Drawing by V8rik, source Wikimedia Commons

The Williamson ether synthesis worked best with alkyl halides that were not prone to (E2) elimination especially when the (Sn2) substitution reaction was favored and the sodium hydroxide was relatively insoluble in the reaction mixture.

Williamson had been a student of Leopold Gmelin (1788–1853)[49] and Liebig. One idea that he supported was that atoms in molecules exchange among molecules. This is, of course, correct for hydrogen-bonding hydrogens and ions, but overall the concept of exchange tended to undermine the idea of "stable" ("inert" would be a better term here) structures. Of course, the idea of general exchange of atoms among stable radicals supported Kolbe and others who resisted the rise of the structural hypothesis in the 1860s.

Reaction of Zn and Na with Organic-halides

We already met Edward Frankland (1825–1899) and discussed his contributions to radical theory. Frankland moved the focus of his research from generating data to support the radical theory to the

---

[49] Gmelin was an inorganic chemist.

study of a new class of organic reagent with his observations on July 12, 1849. He described these events in 1877 and Dietmar Seyferth included the following quote in his summary of organo-zinc chemistry in 2001[50]:

> *"Zincmethyl and zincethyl were the first of these bodies with which I became aquainted: They were discovered on July 12, 1849, in the laboratory of Professor Bunsen in Marburg, during my work on the isolation of the organic radicals. After making the reaction for the isolation of methyl [radical] by digesting methylic iodide with zinc and after discharging the gases, I cut off the upper part of the tube in order to try the action of water upon the solid residue. On pouring a few drops of water upon the residue, a green-blue flame, several feet long, shot out of the tube, causing great excitement among those present. Professor Bunsen, who had suffered from arsenical poisoning during his research on cacodyl, suggested that the spontaneously inflammable body, which diffused an abominable odor through the laboratory, was that terrible compound, which might have been formed by arsenic present as an impurity[51]*

---

[50] D. Seyferth. Zinc alkyls, Edward Frankland, and the beginnings of main-group organometallic chemistry. *Organometallics* 20:2940-2955 (2001).

[51] Indeed, many alloys of zinc contain significant arsenic impurity and any process that digest large quantities of zinc in

*in the zinc used in the reaction, and that I might be
already irrecoverably poisoned. These forebodings
were, however, quelled in a few minutes by an
examination of the black stain [which was zinc] left
upon porcelain by the flame; nevertheless, I did
afterward experience some symptoms of zinc
poisoning."*[52]

His work also included some work with elemental phosphorus and he generalized his ideas based on the concept of organic radicals forming families of analogous compounds based on certain types (ZnH_2[53], PH_3, AsH_3 and SbH_3) with various organic radicals.[54]

---

acid or base my release acutely toxic levels of arsine AsH_3. Unlike most arsenic toxins, arsine attacks the red blood cells and depletes their oxygen carrying capacity (immediate death). The debris from the broken blood cells clogs the kidneys and may cause a lingering death unless the victim is given dialysis.

[52] I enjoy this quote because it captures the instant and uncertainty of discover often found in chemistry.

[53] There was still uncertainty about the atomic weight of hydrogen, hence some molecular formula should have more hydrogens.

[54] One of the compounds predicted by Frankland was trimethylstibine ($(CH_3)_3Sb$), which I prepared using the Grignard reagent in the laboratory of Professor G. G. Long (NC State University) in 1966. I isolated this compound by distillation of the ether followed by distillation of the trimethylstibine from a

By 1850, he had published this work in articles entitled "On a new series of organic bodies containing metals and phosphorus" in which he noted the extreme reactivity of alkyl zinc to moisture and oxygen. He also observed that if diethyl ether were included in the reaction mixture, decomposition of the product was minimized. Frankland pointed out that the affinity of zinc for oxygen would likely allow these compounds to displace oxygen and halogens with alkyl radicals and make new organic compounds through that method. The use of diethyl ether became standardized by 1851 to obtain almost quantitative yields of diethylzinc.[55]

---

wad of magnesium salts that were apparently partially solvated with ether. When the distillation was complete as far as we could tell, the ether solution of the stibine was chlorinated to form the dichloride which was isolated and recrystallized without incident. But, as I was tearing down the original reaction flask (2 L / three-neck with overhead stirring, as I recall) addition of a small amount of water released a large amount of heat, which boiled off the ether that was displaced. There was no fire and for the most part the foam was contained in the hood. But there was enough excitement that the physical chemist (Dr. Moreland) who was in the same room tending his NMR spectrometer, excitedly yelled "kill it, kill it!" To which I answered, "I think I just did."

[55] Soon after Frankland's publications began appearing, Carl Lowig disputed his claims of priority by pointing to some of his research with antimony in the 1840s. After some rancorous

Note that through 1855, the confusion about the atomic weight of hydrogen caused Frankland to give the empirical formula as C_2H_5Zn; but one of the impacts of having a volatile compound with an element as heavy as zinc (atomic mass 65.4) allow very accurate gas densities to be determined. On March 15, 1855, Frankland read his second memoir on "zincethyle" and made the following comments:[56]

> "The vapour volume of zincethyle is highly remarkable, and almost compels us to conclude that the vapour volume of the double atom of zinc is only equal to that of oxygen, instead of corresponding with that of hydrogen, in accordance with the generally received supposition. Zincethyle, therefore, appears to belong to the so-called water type, and consists of two volumes of ethyle and one volume of zinc vapor…."

Clearly, something was wrong with the table of molecular weights.

In the next years, Frankland and other chemists extended his work to compounds of tin, mercury, lead,

---

exchanges Frankland was able to show that Lowig's compounds were not isolated and characterized although they perhaps had been prepared *in situ*.

[56] Quoted from *The Chemical Gazette: Or, Journal of Practical Chemistry*, Volume 13 edited by William Francis, Henry Croft.

boron, aluminum (in 1865 by G.B. Buckton and W. Odling) and beryllium.

In the same timeframe that Frankland was studying zinc alkyls (i.e., 1855), Charles Wurtz (1817–1884) published his observations that sodium metal could be used to couple alkyl radicals obtained from alkyl halides:

$$2R-X + 2Na \rightarrow R-R + 2NaX$$

Alkyl and aryl radicals could be coupled in a similar reaction call the Wurtz–Fittig reaction. This reaction is rarely useful because most functional groups are destroyed by sodium.

The Concept of Valence

One of the unexpected fruits of Frankland's work with organometallics was the observation that the metals formed unique numbers of bonds. This became apparent in the fact that many of the compounds he dealt with were volatile (thus molecular) and retained characteristic combining ratios with the organic radicals. This was a step beyond the combining ratios of Dalton since the organic radicals were known to have exactly one combining unit (e.g., R-). The idea of a directional bond of defined length was not yet accepted but in 1877 Frankland wrote (again quoting from Seyferth, 2001):

"It was evident that the atoms of zinc, tin, arsenic, antimony, &c. had only room, so to speak, for the attachment of a fixed and definite number of the atoms of other elements, or, as I should now [1877] express it, of the bonds to other elements. This hypothesis, which was communicated to the Royal Society on May 10, 1852, constituted the basis of what has since been called the doctrine of atomicity or equivalence of elements; and it was so far as I am aware, the first announcement of that doctrine."

The idea of valence helped pave the way for the structural theory of chemistry and the periodic table (1860-71). From his work in the 1850s, Frankland was closing in on the idea that carbon was tetravalent and he started to view radicals as organic molecules one or two valences less than needed to saturate their maximum valence. About this time (1857), Frankland and Kolbe had a falling out over authorship of an unrelated paper. And, there appears to have been a break in their concepts of radicals. For example, in a letter to Playfair (1860) Frankland writes:[57]

---

[57] Quoted from Edward Frankland: Chemistry, Controversy and Conspiracy in Victorian England

"So-called radicals are nothing more than bodies one or two stages short of chemical saturation."

The impression I have is that Franklin had tired of ignoring the various inconsistences created by Kolbe's strict interpretation of "radicals."

The Structural Theory of Organic Chemistry

Kolbe's world was about to be shaken. In 1857[58], August Kekulé (1829–1896) proposed that carbon atoms normally had a valence of four (i.e., four bonds) and in 1858 he proposed that one of the interesting properties of carbon was its ability to bond to itself.[59]

By Colin A. Russell, Cambridge University Press, 1996. p. 120.

[58] A. Kekulé. Über die s. g. gepaarten Verbindungen und die Theorie der mehratomigen Radicale. *Annalen der Chemie und Pharmacie 104* (2): 129–150 (1857).

[59] It turns out that Archibald Scott Couper (1831–1892) who was working with Charles Adolphe Wurtz (1817–1884) had the same idea in 1858 and might have published it earlier than Kekulé except for some issue with Wurtz. When priority for the idea went to Kekulé, Couper apparently complained to Wurtz and was dismissed from Wurtz's laboratory. Considering that Couper soon suffered a mental breakdown and spend most of the next 30 years in the care of his mother, it seems Couper was probably not reasonable and that Wurtz made a good decision.

This "self-bonding" was somewhat alien to the idea of (non-specific) ionic attraction (affinity among oppositely charged atoms) that Kolbe and every chemist since the work of Charles-Augustin de Coulomb (1736–1806) had held. Moreover, the theory of electrostatic bonding had no provision for preferred directionality (i.e., stereochemistry).

Josef Loschmidt (1821–1895), who is most famous for an accurate estimation of the number of molecules in a volume of gas (1865) that preceded the determination of Avogadro's number by decades, presented two-dimensional formula for organic compounds in 1861:

nylcyanamid ;ch. 238, und beim Einwirken von Chlorcyanur $C_3N_3Cl_3$ auf zwei Äquivalente Anilin das Diphenylchlor-

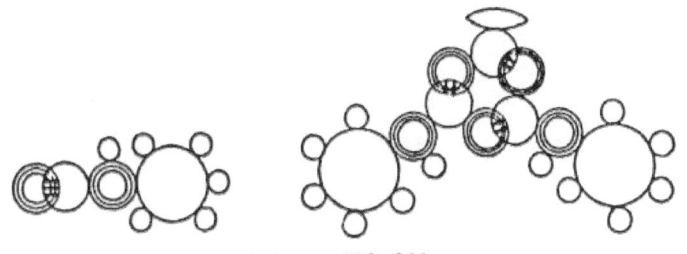

Schema 238, 239.

cyanurdiamin[154] $C_{15}N_5H_{12}Cl$, Sch. 239. Wirkt hingegen Chlorcyan auf zwei Äquivalente Anilin, so erhält man Mola-

---

Couper, however, deserves credit for representing molecular structures with lines representing bonds.

But, his representation of the phenyl group was just a large circle with 5 small circles representing hydrogen atoms (see above) and a larger circle representing an attached heteroatom. Any claim that he was trying to represent benzene as a cyclic compound is laid to rest by the fact that he used similar circles to represent individual atoms or groups of atoms.[60] Indeed, Loschmidt's drawings are actually a representation of Kolbe's electrostatic model of radicals.

Apparently, the first person to think of organic molecules in three dimensions was Alexander Butlerov (1828–1886)[61] who proposed tetrahedral carbon in 1862. Although a tetrahedral arrangement of atoms around a central atom of opposite charge can be rationalized purely on electrostatics (e.g., methane), the idea that carbon was tetrahedral regardless of the radical (i.e., positive or negative) attached to it was as revolutionary to chemistry as quantum theory was to physics; and it would not make sense to the physicists until the work

---

[60] J. Loschmidt. Konstitutions-Formeln der organischen chemie in graphischer darstellung. *Ostwalds Klassiker der exakten wissenschaften* Nr. 190. Source H. S. Rzepa, May 6, 2005.

[61] Butlerov studied with Kekulé and Couper in 1858.

of Erwin Schrodinger (1887-1961) in 1926.[62] Not only did Butlerov assume carbon could be tetrahedral, he drew structures with localized bonds. This led directly to the prediction of structural isomerism. The first breakthrough in this area was the prediction that there should be four isomers of valeric acid ($C_5H_{10}O_2$)

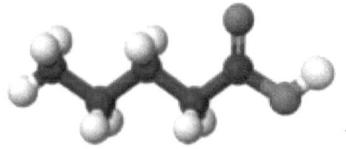

Valeric acid

Drawing by Kemikungen, source Wikimedia Commons

The structure of benzene (empirical formula CH, but Dumas mass indicating C_6H_6) had been a problem of some discussion. But there is no credible claim of a reasonable structure before Kekulé in 1865. It is likely that he was influenced by Butlerov and (allegedly in a dream) Kekulé envisioned benzene (empirical formula CH) being a six-carbon ring, which could be rationalized by alternating C=C double bonds. Kekulé

---

[62] Until the acceptance of wave mechanics describing the electron density around atoms in terms of spherical harmonics, there was no basis for assuming any preferred direction for bonds or fixed bond angles. Kolbe viewed organic radicals as preferred agglomerations of atoms held together like a pile of marbles that had one or more stable overall configuration resulting from electrostatic effects.

even provided what we would call resonance structures:

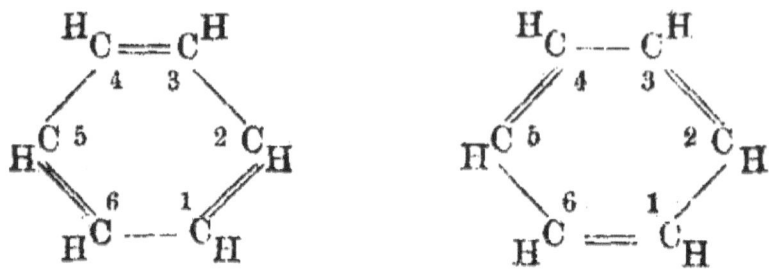

Friedrich August Kekulé von Stradonitz (1829–1896), source Kekulé_-_Ueber_einige_Condensationsproducte_des_Aldehyds.pdf via Wickimedia.

With this idea on the table (1865), a tremendous amount of dye chemistry suddenly made sense. Coupling Butlerov's tetrahedral carbon and double bonds with Kekulé's benzene ring started making chemical reactions and mechanisms understandable. This did not stop Kolbe from denouncing the idea, as it lacked any theoretical basis.

While Runge isolated quinoline from coal tar in 1834, pyridine was not identified until Thomas Anderson (1819–1874) isolated it by heating animal bones in 1849. Wilhelm Körner (1839-1925), a student of Kekulé, proposed that pyridine was an analogue of benzene in 1869.

Drawing by pulkocitron, source Wikimedia Commons

Empirical Rules of Organic Chemistry

It is appropriate that we wind down the empirical period of organic chemistry with two of the best-known empirical rules.[63] Vladimir Markovnikov (1837-1904) and Alexander Zaitsev[64] (1841-1910) both started out as students studying economics with plans for careers in business. But, at Kazan University in Tsarist Russia they encountered the brilliant personality of Alexander Mikhailovich Butlerov and became iconic organic chemists. The two men had very different personalities and developed different views of chemical structure.

Markovnikov began working with Butlerov in 1860 and became an ardent student of Butlerov's ideas of

---

[63] D.E. Lewis. Feuding rule makers: Markovnikov (1837-1904) and Alexander Zaitsev (1841-1910): a commentary on the origins of Zaitsev's rule. *Bull. Hist. Chem.* 35(2):115-124 (2010).

D. E. Lewis, "Aleksandr Mikhailovich Zaitsev: Markovnikov's conservative contemporary," *Bull. Hist. Chem.* 17/18:21-30 (1995).

[64]also Saytzeff or Saytzev.

structural chemistry. Zaitsev, who had recently inherited part of his father's estate, obtained his degree (economics) in 1863 and elected to study chemistry with Kolbe in Leipzig, where he proved his experimental skill by discovering sulfoxides and sulfoxomium salts. Obviously, the two young men were going to obtain very different views of organic chemistry.

In 1865, Markovnikov wrote his master's thesis reflecting Butlerov's views "On the isomers of organic compounds." Ironically, Butlerov sent him to study with Kolbe in Leipzig (1865-66). Zaitsev apparently decided to re-enter the Russian academic system and also returned to Kazan in 1866 as an unpaid assistant to Butlerov because he lacked the Russian master's thesis.

In 1869, Markovnikov wrote his doctoral thesis, which contained the essence of his rules for addition of HX to olefins, i.e., the hydrogen goes on the least substituted carbon. At the same time, Zaitsev got Kolbe to issue him a German doctorate based on his previous work. In all likelihood, Markovnikov and Zaitsev reflected the fundamental clash of Butlerov and Kolbe; and Markovnikov was, no doubt, displeased when Zaitsev received a promotion that Markovnikov would have preferred go to one of his friends. Markovnikov soon left for Odessa (1875) and then to Moscow (1877).

Meanwhile, Zaitsev made his discoveries; published his rule regarding the favored direction of elimination reactions (e.g., dehydrohalogenation) in 1875;[65] and he remained at Kazan.

The Favorskii Rearrangement

Alexey Favorskii (1860–1945) worked with Zaitsev starting in 1882. He discovered a novel rearrangement (1887) that is named for him (Favorskii, A. E. *J. Russ. Phys. Chem. Soc.* 1894, 26, 590), which involves transient formation of a three-membered ring (see below). Some similar reactions appear to go through a different mechanism and are called pseudo-Favorskii rearrangements, which involve a conventional tetrahedral (hydroxide addition to the ketone) intermediate followed by collapse to a carboxyl group

---

[65] A. Saytzeff, Zur Kenntnis der Reihenfolge der Anlagerung und Ausscheidung der Jodwasserstoffelemente in organischen Verbindungen. [Note the order of attachment and elimination of hydroiodic acid in organic compounds]. *Ann. Chem. Pharm.*179:296-301 (1875).

with migration of the primary carbon to displace the chloride by an Sn2 mechanism.[66]

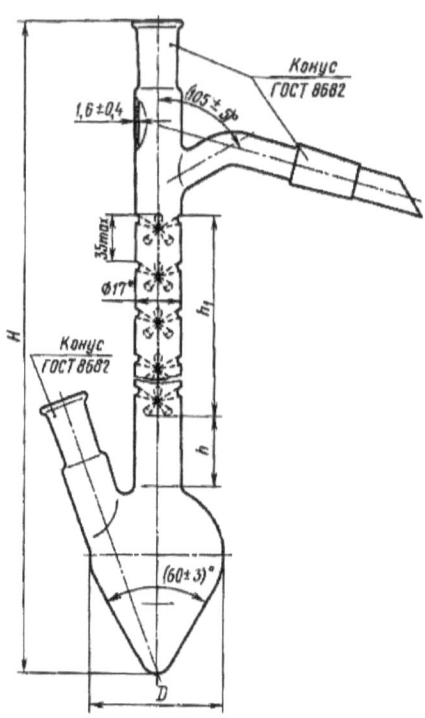

Favorskii Flask

Former Soviet Union, No copyright; source Wikimedia Commons

---

[66]I think there is also a possible carbene reaction mechanism with elimination of both H and Cl from the same carbon followed by insertion in the other alpha C-H bond.

The Favorskii Rearrangement

Drawing by Puppy8800, source Wikimedia Commons

The Beckmann Rearrangement and Caprolactam

Ernst Beckmann (1853–1923) came from a dye stuff-manufacturing family and had enormous practical experience in chemistry before he started university at the age of 22. Following Zaitsev, Beckmann received his doctorate in 1878 from Hermann Kolbe in Leipzig making dialkyl sulfones by oxidation of the corresponding sulfides:

$$R_2S + Permanganate \rightarrow R_2SO$$

After working in other laboratories, he returned to Kolbe; and in 1886, he was making menthone from menthol and experimenting with other terpenoids.

Source: L. T. Sandborn. Menthone. *Org. Synth.* 9:52 (1929).

He used hydroxylamine (H_2NOH) to make oximes from his ketones and discovered that treatment of these with strong acid resulted in a complex rearrangement to yield an amide. For example, with cyclohexanone he produced caprolactam[67]:

cyclohexanone   cyclohexanoxime   caprolactam

Drawing by V8rik, source Wikimedia Common

---

[67] Compare this to the Baeyer–Villiger Oxidation/Rearrangement.

Beckmann's Thermometer and Molecular Weighs

Thanks to the work of Dumas, relative molecular weights of volatile compounds were routinely available in the late 1800s.[68] This was not the case with non-volatile compounds. To help resolve questions about the nature of oximes, which were starting to play a major role in resolving chemical structures and in general synthesis, Beckmann introduced the technique of freezing point depression. In 1788, Charles Blagden (1748–1820), an assistant to assistant to Henry Cavendish, had noted that the depression of freezing points was related to the amount (mass) of solute dissolved in the solvent (i.e., Blagden's law). With the aid of a thermometer designed to very accurately measure small changes in temperature in a narrow range, Beckmann developed

$$\Delta T_f = m$$
$$K_f$$

---

[68] The problem with diatomic gases still existed and caused some confusion.

techniques for measuring freezing point depression and boiling point elevation (ebulliometry, e.g., Raoult's Law of 1871) and relate these to the molecular weight of the solute. These techniques were expanded and perfected between 1888 and the 1920s.

Source: Ervin Sidney Ferry. Page 67 of
A Handbook of Physics Measurements
(1918) Copyright expired. Wikimedia Commons.

Adolf von Baeyer (1835–1917)

It is most fitting that we end the "ascent of the structural theory" with brief mention of Adolf von Baeyer. Baeyer [pronounced "buyer' in English] was a student of Robert Bunsen and August Kekulé who initiated the "structural movement" (1858). Indeed, it is hard to imagine that Baeyer had no influence on Kekulé's reasoning. After Justus von Liebig died in 1873, Baeyer was selected as his successor at the University of Munich.

Baeyer's personal contributions to organic chemistry were relatively subdued. He synthetized phenolphthalein (from phenol and phthalic acid) in

1871 and indigo in 1882 (from o-nitrobenzaldehyde and acetone)[69]:

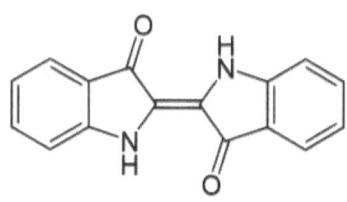

phenolphthalein
Drawing by Benjah-bmm27,
source Wikimedia Commons

indigo
Drawing by Yikrazuul,
source Wikimedia Commons

He also was one of the first to consider strain energy of distorted bonds. Nonetheless, his major contribution was likely the series of productive German chemists who were trained in his laboratory in the theory of structural chemistry.

---

[69] This reaction (aldol condensation) is now recognized as a Claisen condensation after one of his students.

III. Classical Organic Chemistry (1870-1930)

As the preceding part suggests, there was a substantial amount of empirical organic chemistry known by 1870. I have chosen this as the beginning of classical organic chemistry because at this point the debated over atoms, molecules and heat were largely resolved (at least for chemists, though not entirely for physicists) and the notion of vital force had been put to bed. Structural chemistry was becoming popular in spite of its lack of direct theoretical basis.

These facts more or less coincided with the commercial availability of petroleum,[70] which was a great new resource of raw material for organic chemistry. The economic value of petroleum created a major incentive to develop chemistry depending on this raw material. Organic chemistry, thus, developed as a new science. In the period 1900-1930, an interest in the biological functions of molecules was initiated. World War I

---

[70] Remember Drake's well was successful in 1859 and others soon followed.

(1914-1918) diverted many efforts to munitions manufacture and general technological developments in internal combustion engines that drove petroleum-based research. But, without a clear understanding of atomic electronic structure and bonding, there was still a large element of empiricism in organic chemistry until the 1930s.

1. Standard Oil

As we saw earlier, liquid fuels had great convenience and customer-appeal. Whale oil and olive oil were in short and declining availability in the early 1800s and coal-derived kerosene became popular as the only alternative. Unfortunately, the method of production of kerosene (which was operationally defined, not chemically defined) produced a lot of lower molecular weight flammable compounds and repeated distillations were required to remove these components that made kerosene dangerously flammable in many of the lamps used at the time. There were no legal or ethical standards and the quality and danger of kerosene varied widely. In this market (*circa* 1865), John D. Rockefeller had a good idea: He standardized his kerosene calling it "standard oil" and advertised its safety. He was able to do that because he started using petroleum, which came available in 1860 (following Drake's well) as his only source of material and it was easier to distill without cracking to smaller molecules. Standard Oil was incorporated as a company in 1870 in Cleveland, Ohio. Rockefeller then proved himself to be a smart and ruthless capitalist as he systematically eliminated all his small and disorganized competitors. He also tamed the contemporary corporate giants (i.e., the railroads) by building pipelines to transport his raw materials and products. Having established his monopoly in the refining and distribution areas (a "horizontal trust"), he proceeded to leverage is power into petroleum production and exploration. Along the way he successfully battled financial institutions, the railroads, the steel industry and even the newly introduced electric industry.

We will return to the petroleum industry after the introduction of Henry Ford's Model-T automobile (1908) and anti-trust decision against Rockefeller (May 15, 1911).

2. Accurate Heats of Combustion and Hydrogenation

Major advances in the practice and theory of organic chemistry can often be attributed to new or improved physical methods of analysis or purification. Here we discuss one of the first major factors that helped organize and rationalize organic chemistry: accumulation of accurate heats of combustion for a number of different compounds and clarification of the roles of enthalpy and entropy in thermodynamics.

In about 1864, Marcellin Berthelot followed the work of Joule by devising a "bomb calorimeter" (constant volume) in which samples to be analyzed are placed in a sealed metal container, which is charged with an excess of oxygen. The sample is then ignited by passing a current through a wire and combustion follows rapidly and completely (to carbon dioxide and water) in the oxygen atmosphere. In particular, for organic compounds of C, H and O, the products of combustion are always (only) H_2O and CO_2. This methodology provided a very convenient basis for comparison of the heat derived through combustion.

Based on his observations on heat of combustion and his interest in organic synthesis, Berthelot composed a concept that if a reaction proceeds, it is likely to favor the more stable products (as determined by loss of potential energy to produce heat energy).

With the bomb calorimeter now available, Julius Thomsen (1826-1909) began doing similar calorimetry experiments in 1869. Through general misunderstandings, the ideas of Thomsen on spontaneity and Berthelot on preferred product yields, were conflated into the Thomsen-Berthelot principle, which was generally understood to mean that spontaneous reactions are (always) exothermic and vice versa.[71] In 1873 Berthelot stated:

> "Every chemical change accomplished without the intervention of external energy tends towards the production of a body or system of bodies, which produce the most heat."

---

[71] This principle was generally accepted until 1882 when Hermann von Helmholtz (1821–1894) pointed out that it is only the "free energy" that determines the spontaneity of reactions. This principle is actually true at "absolute zero" but needs to be corrected for energy involved in bringing the reactants and products to the ambient temperature of reaction. See Walther Nernst's (1864-1941) heat theorem 1906.

This principle turns out to often be true, but is in error in the general case. Examples of spontaneous endothermic reactions were well known. Thus, I prefer to think that Berthelot was thinking of a limited scope of reactions of similar organic compounds leading to similar products. Ironically, Thomsen (whose experience predated Clausius's 1865 articulation of entropy) had made similar statements *circa* 1854, and challenged Berthelot's priority of this *erroneous* principle. The history of this episode has been discussed elsewhere.[72] Suffice it to say that Berthelot eventually acknowledged that enthalpy alone does not determine the spontaneity of reactions. The importance of his work is not in development of thermodynamic theory, but rather in development of bond theory. His data started gaining substantial attention after 1883 when the details of his apparatus were published.[73]

---

[72] R. G. A. Dolby. Thermochemistry versus Thermodynamics: The Nineteenth Century Controversy. *History of Science*, Vol. 22, p.375-400.

[73] E.S. Domalski. Selected values of heats of combustion of organic compounds containing the elements C, H, N, O, P and S. J Phys Chem Ref Data. 1(2):221-277 (1972).

In the 1880s, Samuel Wilson Parr (1857-1931) became interested in the heating value of fuels. After passing through a number of designs for calorimeters, his work made the "Parr Bomb" the standard in the field. In principle, Parr's calorimeter was very similar to the apparatus used by Berthelot.

Cotton Seed Oil and Hydrogenation

The invention of the cotton gin (1794) facilitated rapid separation of cotton fibers from the cotton seeds. For decades the cotton seeds piled up and were discarded as waste. In 1857, William Fee invented a machine that would efficiently crush the tuff seeds and separate the hulls from the meat from which oil could be pressed. The invention came too late for cotton seed oil to be useful as fuel as petroleum was widely available by 1870. The oil was, thus, surreptitiously used to dilute imported olive oil and by American meat packers to fortify lard and other animal fats. Ironically, the scheme was detected in 1884 when Armour and Company tried to monopolize the lard market and discovered that there was more lard on the market than could be produced by all the hogs in the US. The US Congress got involved and passed regulations that fortified Lard must be labeled as "lard compound."[74] And olive oil producers banned the practice (1883).

Cotton seed oil became worthless again and was picked up by Procter & Gamble for the manufacture of candles. But, the market of candles was declining with the advent of kerosene and electricity.

Meanwhile, James F. Boyce had developed a method for hydrogenating vegetable oils while working on soap products for the N.K. Fairbank Co. in Chicago.

In France, Paul Sabatier (1854–1941) was a student of Berthelot (1880) with interest in catalysis. He developed the principle that an effective heterogeneous catalyst should bond with the substrate, but not so strongly that it blocked the catalyst. Therefore, he worked with hydrocarbons in the gas phase. He found that nickel was more effective than platinum in hydrogenation of olefins (1897) and started collecting data on the heat of hydrogenation of unsaturated organic compounds. For this, he shared the 1912 Nobel Prize[75] with Victor Grignard:

---

[74] O. P. Snyder. *Food law in the United States* pp 12-29 (1997).

[75] Nobel Lecture, December 11, 1912: The Method of Direct Hydrogenation by Catalysis

Decisive success came at the end of 1900 when, with Senderens, I found that benzene can be totally changed into cyclohexane in contact with nickel at a temperature of about 180°C. After that I was absolutely confident of the general nature of the method, the principle of which we stated at the beginning of 1901: "Vapour of the substance together with an excess of hydrogen is directed on to freshly reduced nickel held at a suitable temperature (generally between 150 and 200°C)."

In Germany, Wilhelm Normann (1870–1939) followed Sabatier's work and put it to practical use in the hydrogenation of liquid oleic acid to solid stearic acid. He obtained patents in Germany and Britain by 1903 and these patents were the object of commercial competition among several companies in Europe. Ultimately, Edwin C. Kayser a chemist with Joseph Crosfield and Sons moved to Procter & Gamble with the idea of hydrogenating cotton seed oil as an edible product. They succeeded and the product was marketed as Crisco™ shortening (1911).

In 1924, Murray Raney (1885–1966) discovered that an aluminum-nickel alloy can be leached with base to remove the aluminum leaving finely divided nickel suitable as a catalyst. In 1926, Parr introduced its shaker hydrogenator based on Voorhees and Adams.[76] Heats of hydrogenation made comparison of the stability of olefins and aromatic compounds straightforward and ultimately led to the theory of aromatic stabilization.

3. Some Well-Known Reactions from the late 1800s

Friedel-Crafts Alkylation and Acylation

Charles Friedel (1832–1899) was a student of Louis Pasture and became a chemistry professor in 1876. James Mason Crafts (1839-1917) was an American chemist who first met Friedel in 1861 on a trip to Europe and then returned to Paris (1870s) where he collaborated with Friedel.   They published two articles in 1877 describing what is now known as Friedel-Crafts alkylation and acylation. Some of what we know has been added by other chemists over the years, but briefly, the reaction is an electrophilic aromatic

---

[76] V. Voorhees and R. Adams. The Use of the Oxides of Platinum for the Catalytic Reduction of Organic Compounds. *J. Amer. Chem. Soc.* 44 (6):1397 (1922).

substitution (i.e., an electrophile attacks a benzene ring displacing a hydrogen, H^+). The electrophile is typically formed in a nonpolar solvent by the action of a Lewis base ($AlCl_3$ or $FeCl_3$) with an alkyl chloride or acetyl chloride. In the case of secondary or tertiary alkyl chlorides, typical carbocation rearrangements are observed implying that an essentially free carbocation is formed (although it is probably an ion-pair) with primary alkyl chlorides the halogen-carbon bond is not likely broken (although it is strongly polarized). In the case of acyl chlorides ($RCOCl$), the reaction mechanism is generally written as involving an acylium carbocation ($R\text{-}CO^+ \leftrightarrow R\text{-}^+C=O$).[77]

Since alkylation of an aromatic increases its nucleophilicity, Friedel-Crafts alkylation notoriously produces multiple substitutions (even when the electrophilic reagent is added to an excess of the aromatic). On the other hand, acylation deactivates the ring to further electrophilic substitution and mono-acylation reactions are readily achievable. The acyl group can then be reduced to an ethyl group or

---

[77] Nitriles can be made to alkylate aromatics with $AlCl_3$ (see the Gattermann aldehyde synthesis and the Gattermann–Koch reaction).

modified by Grignard reaction to other aromatic derivatives.

Diazo Chemistry: ethyl diazoacetate

Eduard Buchner (1860-1917)[78] worked with Adolf Baeyer [von Baeyer] (1835–1917) and Emil Fischer (1852–1919) before receiving his doctorate in 1888. But in 1885[79], he published a paper with Theodor Curtius (1857–1928)[80] on the synthesis and reaction of ethyl diazoacetate from ethyl glycinate hydrochloride. Subsequently, he discovered the Buchner ring expansion reaction:

bicyclo[4.1.0]heptadiene   cycloheptatriene

Drawn by PLTL09, source Wikimedia Commons.

---

[78] Younger brother of Hans Ernst August Buchner (1850–1902).

[79] E. Buchner and T. Curtius. *Ueber die Einwirkung von Diazoessigäther auf aromatisch Kohlenwasserstoffe. Ber. Dtsch. Chem. Ges.* 18: 2377 (1885).

[80] Noted for the reaction in which and acyl azide rearranges to an isocyanate (Curtius rearrangement, 1890).

The Michael Reaction

Arthur Michael (1853–1942) had a long and distinguished career. He was fortunate to be born into a wealthy family in New York and was too young for the American Civil War. His early education was by tutors in a private laboratory. In 1871, he traveled to Europe and studied under the most eminent chemists (Hofmann, Bunsen, Wurtz and Mendeleev) in Germany, France and Russia. When he returned to the US (1881); although he had no formal college degree, he became a professor of chemistry at Tufts University; married a student; and moved to the Isle of Wright where he set up a private laboratory (1890).

Apparently, he did his most memorable work at Tufts. For in 1887 he published a paper describing the Michael Reaction.[81]

Drawing by Dissolution, source Wikimedia Commons

---

[81] A. Michael. *Ueber die Addition von Natriumacetessig- und Natriummalonsäureäthern zu den Aethern ungesättigter Säuren.* *Journal für Praktische Chemie* 35:349–356 (1887).

This work follows from an attempt to repeat the synthesis of a cyclopropane derivative.[82] In the prior art, the alpha-beta unsaturated ketone was apparently generated *in situ* by elimination from the corresponding dibromide. That same year, Ludwig Claisen (1851–1930)[83] claimed precedent from work done earlier, but he had merely observed some byproducts that he did not follow up. Michael's work explained Claisen's observations.

Michael returned to the US and taught at Tufts (1894-1907), retired, and then taught at Harvard (1912-1936).

The Claisen and Dieckmann Condensations

Ludwig Claisen (1851–1930) and Walter Dieckmann (1869–1925) are representative of the German chemists who studied in Adolf Baeyer's laboratory. They studied the condensation of carbanions (enols) produced by treatment of ketones with base. The reactions are very similar but the Claisen condensation is intermolecular

---

[82] M. Conrad. M. Guthzeit. *Ueber die Einwirkung von α-β-Dibrompropionsäure auf Malonsäureester"*. *Berichte der Deutschen Chemischen Gesellschaft* 17(1):1185–1188 (1884).

[83] L. Claisen. *Bemerkung Über die Addition von Aethylmalonat an Körper mit doppelter Kohlenstoffbindung"*. *Journal für Praktische Chemie* 35(1):413–415 (1887).

whereas the Dieckmann condensation in intramolecular:

Claisen Condensation (1883-1887)

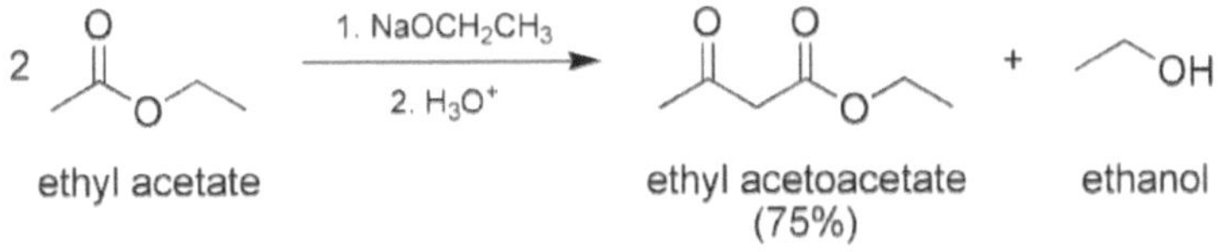

ethyl acetate   ethyl acetoacetate   ethanol
                                        (75%)

Drawing by Jesse, source Wikimedia Commons

Dieckmann Condensation (1894-1901)

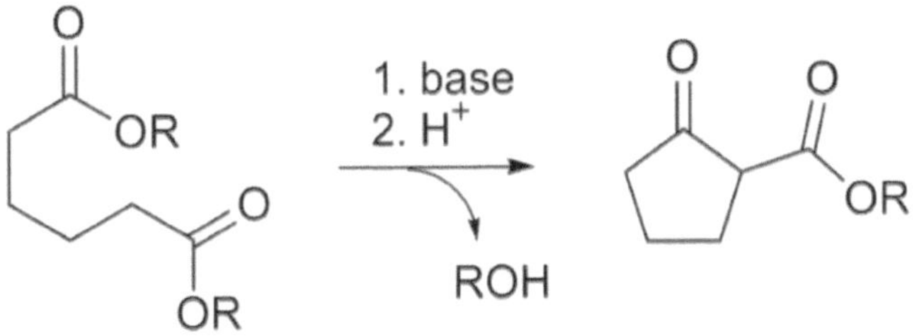

1. base
2. H$^+$

ROH

Drawings by Choij, source Wikimedia Commons

Baeyer–Villiger Oxidation/Rearrangement

Victor Villiger (1868–1934) was another one of von Baeyer's students at Munich. They perfected the attack of a peroxide or peroxyacid (e.g., monoperoxysulfuric acid) on a ketone to yield (after a shift of a carbon to the attacking oxygen) an ester. When the carbonyl is part of a ring, the product is a lactone. This reaction stirred a substantial interest in the mechanism since the transferred group retains its absolute configuration. Over the years various mechanisms were proposed, but the question was not resolved until 1953 with the use of ^{18}O-labeling on the ketone.[84] The mechanism appears to involve a typical tetrahedral intermediate after addition of the peroxide to the ketone, followed by re-formation of the ketone with shift to one of the carbon groups to the electrophilic oxygen created by displacement of the acid anion.

$$[RR'C(O^-)(O\text{-}Oacid)] \rightarrow RCOOR' + {}^-Oacid$$

---

[84] W. von E. Doering, E. Dorfman. Mechanism of the peracid ketone-ester conversion. Analysis of organic compounds for oxygen-18. *J. Am. Chem. Soc.* 75(22):5595–5598 (1953).

4. Optical Activity and the Structural Theory

Kolbe Makes a Fool of Himself

Although the idea of tetrahedral carbon had been floated and individual molecules were known to rotate the plane of polarized light, it was not until 1874 that these two ideas came together. In that year Joseph Le Bel (1847-1930) and Jacobus Van't Hoff (1852-1911)[85] independently proposed that the arrangement of atoms in space around a tetravalent carbon atom was not fluid or arbitrary…it was actually fixed such that C(ABDE) could be isolated from its mirror image and the two (optical) isomers rotated light in opposite directions although in respect to all their physical properties they were identical. From this concept, they could calculate the number of optical isomers of compounds with multiple chiral centers: 2^n where n = the number of asymmetric carbons.

Naturally, Kolbe denounced these ideas viciously. It could be argued that van't Hoff and Le Bel were

---

[85] J. H. van't Hoff (1874) *Die Lagerung der Atome im Raume (The Arrangement of Atoms in Space)* Translated by F. Herrmann 1877

building castles on a foundation of sand (i.e., nothing was really known about the structure of radicals or covalent chemical bonds), but Kolbe's criticism has gone down as one of the most foolish and arrogant personal attacks in the chemical literature (*Journal für praktische Chemie*, 15, 473):

> "…This natural philosophy, which had been put aside by exact science, is at present being dragged out by pseudoscientists from the junk-room which harbors such failings of the human mind, and is dressed up in modern fashion and rouged freshly like a whore whom one tries to smuggle into good society where she does not belong."

> "Whoever considers this apprehension to be exaggerated should read, if he can manage it, the recently published pamphlet, "*The arrangement of atoms in space*", by Messrs. van't Hoff and Herrmann, which teems with fantastic trifles."

> "…A J. H. van't Hoff who is employed at the Veterinary School in Utrecht appears to find exact chemical research not to his taste. He deems it more convenient to mount Pegasus (evidently loaned from the Veterinary School) and to proclaim in his "La chimie dans l'espace" how, to him on the chemical Parnassus which he ascended in his daring flight, the atoms appeared to be arranged in the Universe."

"It is characteristic of today's uncritical and criticism-hating time, that two virtually unknown chemists, the one from a veterinary school, the other from an agricultural institute, judge the most profound problems of chemistry which probably will never be answered. They judge these most important problems, especially the question as to the spatial orientation of the atoms, with a cock-sureness and insolence which can only astound a true student of natural science."[86]

Fortunately, Emil Fischer (1852-1919) was a more practical man.

---

[86] I once received a reviewer's comment that described my work as "much ado about nothing of any real scientific interest." In recent years, I have published some hypotheses, which probably invoke similar thoughts but nothing has appeared in print, yet.

Phenylhydrazones and Indole Synthesis

In 1875, Emil Fischer found that he could reduce phenydiazonium salts to the corresponding hydrazines with sodium sulfite.

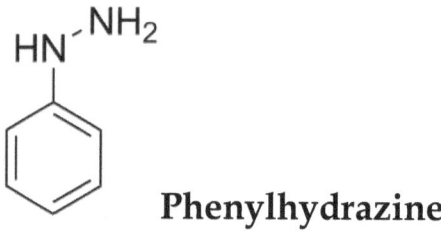

Phenylhydrazine

Drawing by Edgar181, source Wikimedia Commons

In 1877, Fischer discovered that phenylhydrazine reacts with aldehydes and ketones to form compounds that are frequently easy to crystallize with defined melting points and characteristic colors.

In 1883, He found that reaction of phenylhydrazine with aliphatic ketones and aldehydes produced an intermediate phenylhydrazone that could be rearranged into an indole.

Drawing by Nuklear, source Wikimedia Commons

Monosaccharides:
Indirect Proof of the Structural Theory

Sugars in particular were studied extensively in association with the fermentation of carbohydrates $(C_n(H_2O)_{n-1})$ to alcohol and they were often found to be optically active. By 1886, five optically active sugars were known:

Aldohexoses (n = 6)

(+)-glucose (dextrose; from its sign of rotation),

(+)-galactose

Aldopentoses (n = 5)

(-)-fructose (levulose; from its sign of rotation)

(+)-arabinose

(-)-xylose (wood sugar)

These sugars were all derived from natural sources.

By careful chemical analysis, the empirical formula of the compounds could be determined. Emil Fischer extended his phenylhydrazine work to sugars in the mid-1880s and discovered that they would from osazones:

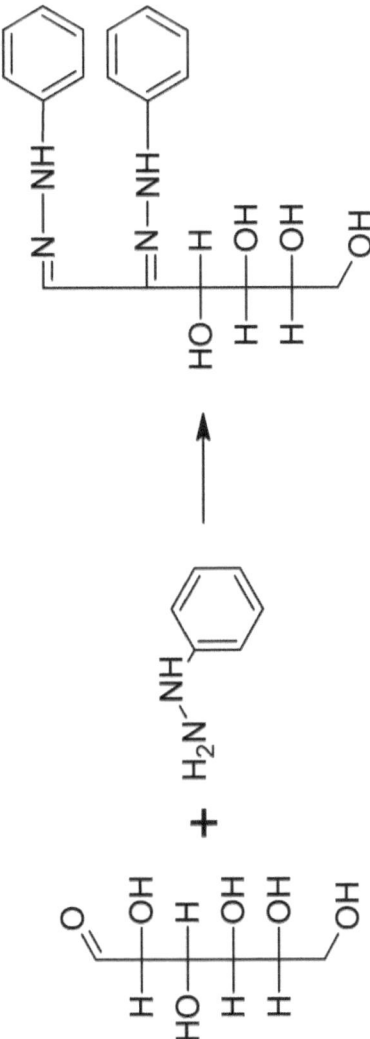

Drawing by Shoyrudude555, source Wikimedia Commons

The osazones were convenient for identification of sugars and got Fischer interested in the problems of structural chemistry. By now, the chemical community had chosen sides between those who believed in the structural theories of Kekulé, van't Hoff and Le Bel and those who did not (led by Kolbe).

A major synthetic breakthrough was made about this time by Heinrich Kiliani (1855-1945). By modifying the methods of Strecker (1850), Kiliani discovered that cyanohydrins could be made from the aldehyde group of these sugars and after hydrolysis (which extended the carbon chain by one unit) cyclic esters (lactones) were formed that could be reduced to racemic mixtures of sugars with one more carbon atom. The lactones could be identified by hydrolysis of the ester linkage and reduction of the alcohol groups with HI/red phosphorous to yield the corresponding aliphatic acid.

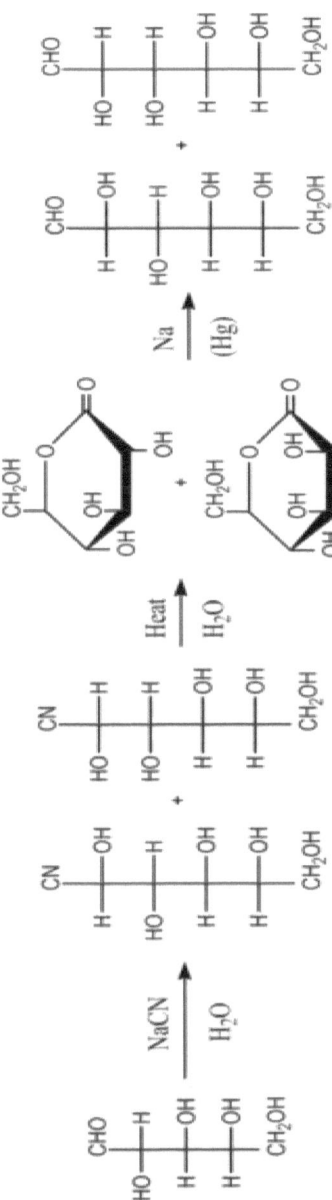

Drawing by Jeff Dahl and revised by DMacks, source Wikimedia Commons

Note that the lactones are diastereomeric with cis- and trans- alcohol groups, which results in substantial physical property and reactivity differences. This feature made their physical separation relatively easy.

With these tools at his disposal, Fischer reasoned that he could unravel the issues associated with the optical activity of sugars. In particular, van't Hoff and Le Bel predicted that the aldohexoses with empirical formula $C_6H_{12}O_6$ should include 16 optically active isomers ($2^4 = 16$):

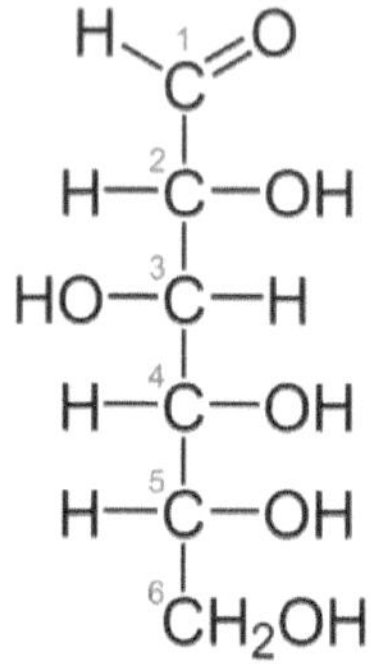

$$
\begin{array}{c}
H-\overset{1}{C}\!=\!O \\
H-\overset{2}{C}-OH \\
HO-\overset{3}{C}-H \\
H-\overset{4}{C}-OH \\
H-\overset{5}{C}-OH \\
\overset{6}{C}H_2OH
\end{array}
$$

Drawing by Yikrazuul, source Wikimedia Commons

The figure above is called the Fischer projection with all the horizontal bonds out of the plane of the page and the vertical bonds into the plane of the page.

By a series of chemical transformations and careful reasoning, Fischer sorted the isomers into groups with equivalent stereochemistry in 1891. But he still did not know the absolute configurations of the two groups.

He ultimately guessed (correctly)[87] for glucose and the configurations for all the other isomers followed from this. Note that Fischer used d/l the same as +/-. [88]

Glucose is more stable in the hemi-acetal form:

α-D-(+)-Glucopyranose

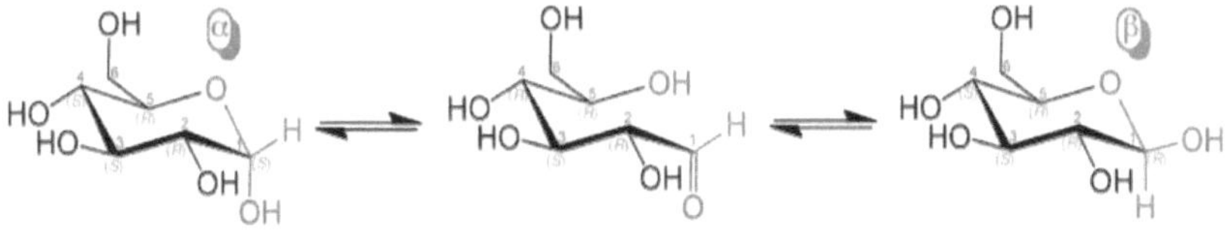

Drawing by Jü, source Wikimedia Commons

Starch

Unlike sugar, starch is a carbohydrate that is found in two forms: Linear-chains of glucose called amylose and branched chains called amylopectin

---

[87] Using Na and Rb tartrate: J.M. Bijvoet et al. Determination of the absolute configuration of compounds with x-rays. *Nature* 158:271 (1951).

[88]The meaning of the d/l nomenclature has led to some confusion and today we use D/L to indicate the absolute configuration at the next to the last carbon (counting down from the top) of chiral monosaccharides and chiral alpha-amino acids.

Drawings by NEUROtiker, source Wikimedia Commons

Animals store glucose as starch (glycogen), which can be broken down as needed to provide a constant level of glucose in the blood, which supplies glucose to the brain and other organs. Plants also store starch.

5. Structural and Steric Effects

The hypothesis of van't Hoff (1874) backed up with its successful application by Fischer (1891) to rationalize the structures of the aldohexoses caused others to think more deeply about the implications of structural chemistry.

Ring Strain

In 1885, Adolf Baeyer (1835-1917) realized that the perfect tetrahedral angle is 109.5 degrees. From this, he immediately deduced that ring compounds that facilitated this angle (e.g., five-membered ring 108 degrees) would be more stable than other structures. Hermann Sachse (1862-?) was unknown to the organic chemist of his day, but made models that showed that tetrahedral carbons could form six-membered (and larger) rings with no ring strain (*Berichte*, 23, 1363-1370 (1890)). He also anticipated that such rings could flip between different conformations (boat and chair) with minimal ring strain. But Baeyer refused to accept the idea that the rings could be non-planar (*Liebig's Annalen*, 258, 145 (1890)). Sachse's attempts to win the hearts of organic chemists were frustrated by the fact that his later papers were written with geometric and/or trigonometric arguments.[89] In 1918 the crystal structure of diamond was determined and it clearly had non-planar six-membered rings. The organic

---

[89] *Zeitschrift für physicalische chemie*, 10:201-241 (1892) and 11:185-219 (1893).

chemists did not fully get onboard until the 1930s with the work of Odd Hassel (1897–1981).[90]

Drawing by Benjah-bmm27, source Wikimedia Commons

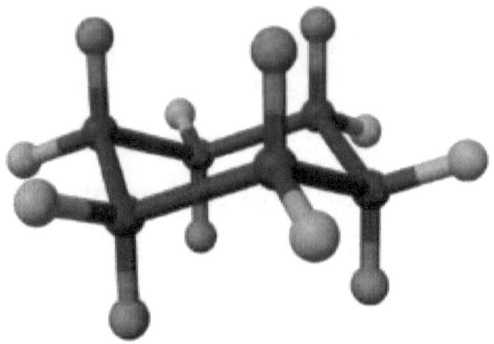

A ball-and-stick model of cyclohexane showing axial position in red and equatorial in blue.

Steric Hindrance

In 1872, Hofmann had observed that pentamethylaniline did not form methyammonium salts. And, there were other examples of ortho-substituents on benzene rings making reactions slow or complete failures. In 1894, Victor Meyer introduced the term "steric hindrance." He showed that silver salts of

---

[90] He received the Nobel Prize for this work in 1969 for work on charge-transfer complexes.

benzoic acids all readily reacted with ethyl iodide, but those with various ortho-substituents were unexpectedly slow to react.[91]

In 1922, 2,2'-dinitrodiphenyl-6,6'-dicarbonylic acid was resolved into optically isomers.[92]

6. The First Synthetic Plastics

Celluloid

Alexander Parkes (1813–1890) who was noted for his work on vulcanization of rubber and metallurgy was the first (1850) to produce celluloid plastics from nitrocellulose. This material took advantage of the natural polymer backbone provided by cellulose and broke the crosslinks by nitration to produce a pliable material that could be mixed with camphor and molded into various shapes (1856). Of course, nitrocellulose was extremely flammable and is degraded overtime by moisture and fungi. During

---

[91] V. Meyer. *Chem. Ber.* 27, 510 (1894).

[92] G.H. Christie and J. Kenner. *J Chem. Soc.* 121:614 (1922).

1866-68, he manufactured a celluloid resin commercially. In 1870, John Hyatt (1837–1920) established a company in the U.S., primarily to manufacture billiard balls. The processes expanded with blow molded shapes and products such as false teeth, combs, mirrors, and finally film for cameras. Soon, he was challenged by Daniel Spill (1832–1887) formerly a partner of Parkes, for infringement of Parkes's patent. The legal proceedings were drawn out. Ultimately, the court ruled that Parkes was the inventor (1884) and Spill died in 1887. In the 1890s celluloid found a market in photographic and movie film, which it retained into the 1950s. Most of the rigid applications of celluloid were replaced with Bakelite after 1910. Today, it is only used for ping-pong balls.

Bakelite

In 1872, Adolf Baeyer noticed that phenol and formaldehyde formed a resin, but he did not pursue the observation. Leo Baekeland (1853-1944) was born into the family of a cobbler and maid in Belgium and trained as a chemist. He began a successful academic career, but on a trip to New York in 1889, he realized the possibility of making serious money from chemical products. He moved to the U.S.; worked for a photographic company for a few years; started his own company making the first photographic paper; became an American citizen in 1897;

and sold his company to George Eastman in 1899 with a net gain of over $200,000 (a fortune at that time).

Now, he was caught up in the entrepreneurial spirit of America in 1900. He realized the limitations and expense of natural rubber and other natural polymers and began experiments with compounds that might react to form polymers. In 1907, he found that phenol and formaldehyde were condensed to form a liquid resin that could be directly molded into solid products or produced as a powder that could be used to mold a wide variety of products (especially electrically insulating materials in the growing industries: lights, phones, cars, airplanes). For the next 30 years, Bakelite was the principal synthetic plastic. His company was sold to Union Carbide in 1939 and in 1944, the world production of Bakelite was about 175,000 tons per year.

Linear Polymers

In the early 1900s, the early plastics were (and were assumed to be) three-dimensional agglomerations that could be cured into rigid shapes by molding. In fact, Bakelite was a reasonable model of this concept. But with three-dimensional polymers it was not possible to make flexible synthetic fibers. Hermann Staudinger (1881-1965) was the first to envision the possibility of linear polymers of repeating monomer units (1920).[93]

---

[93] H. Staudinger. Über Polymerisation. Ber. *Deut. Chem. Ges.* 53(6):1073 (1920).

At the time, this idea was considered unlikely because no one had any idea how to reliably cause the repetitive terminal addition that such a process required. Staudinger received the Nobel Prize in 1953 for his foresight. Some of the first proof for polymerization of monomer units into a repeating structure came from the work on ozonolysis by Carl Harries (1866–1923) with natural rubber.

Ozonolysis of C=C Bonds

Ozone (O_3) was discovered by Christian Schönbein (1799–1868) in 1840. At that time, the structural theory of organic chemistry did not exist; and although he may have realized that reaction of ozone with rubber and terpenes broke the materials into simpler forms, his contribution was limited. Carl Harries, with the advantage of a well-developed structural theory, began a series of experiments in the early 1900s, which led to appreciation that carbon-carbon multiple bonds were cleaved by ozone.[94] The mechanism of this novel reaction was not understood, however, until 1950 when

---

[94] C. Harries. *Ueber die Einwirkung des Ozons auf organische Verbindungen. Liebigs Annalen der Chemie* 343(2–3):311–344 (1905).

Rudolf Criegee (1902-1975)[95] proposed that the ozone adds in a 1,3-manner to the double bond and the 5-membered ring (1,2,3-trioxolane) dissociates into a zwitterion (carbonyl oxide) and ketone, which rotate within a solvent cage and recombine as a stable ozonide (i.e., 1,2,4-trioxolane).[96]

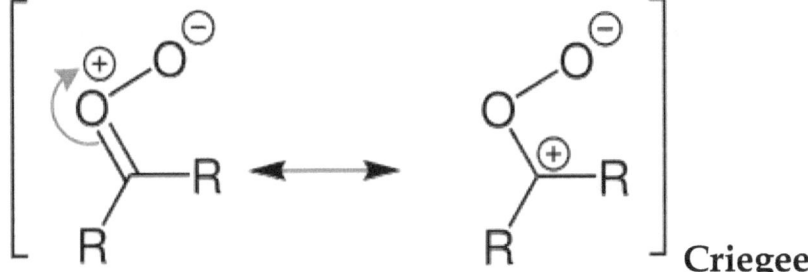

Criegee Intermediate

Drawing by Wickey-nl, source Wikimedia Commons.

The ozonide may be broken down under reducing or oxidizing conditions to produce alcohols, ketones and/or carboxylic acids.

---

[95] Criegee was drafted into the German army in WWII and was seriously wounded on the eastern front in 1942. After the war he played a role in rebuilding the German chemical industry.

[96] R. Criegee. Mechanism of ozonolysis. *Angewandte Chemie* International Edition in English 14 (11): 745–752 (1975).

7. Organometallic Reagents for Synthesis

Reformansky Reaction 1887

Sergey Reformatsky (1860–1934) accumulated a sterling resume studying under Zaitsev (1882), Meyer and Ostwald (1891). In 1887 he published a paper[97] describing a novel synthetic reaction facilitated by an organozinc reagent. Although Frankland had made numerous organozinc compounds (1850-1877), his work had mainly contributed to the understanding of organic radicals, valences and atomic masses. Reformansky opened the idea of using these reagents for organic synthesis.

Drawing by ~K, source Wikimedia Commons

---

[97] S. Reformatsky. *Neue Synthese zweiatomiger einbasischer Säuren aus den Ketonen. Berichte der Deutschen Chemischen Gesellschaft* 20 (1): 1210–1211 (1887).

The reaction shown above has survived as a named reaction, because the reagents of Victor Grignard are too reactive for this particular transformation.

Grignard Reagents 1900

In 1900, Victor Grignard (1871-1935) discovered that magnesium metal would react with alkyl or aryl chlorides, bromides or iodides in the presence of anhydrous diethyl ether to produce an ether-soluble reagent with the formula "RMgX." The reaction occurs on the surface of the metal by successive single electron transfers. The actual composition of the reagent depends somewhat on the anion (X) and the ether. For many years there was a debate concerning the existence and position of the "Schlenk equilibrium."[98, 99]

$$2\ RMgX \rightleftharpoons R_2Mg + MgX_2$$

---

[98] Wilhelm Schlenk (1879–1943) Studied Grignard reagents and discovered organolithium reagents (1917). Along the way, he and his son developed glassware for manipulating air-sensitive compounds.

[99] E.C. Ashby, G. Parris and Frank Walker. Direct nuclear magnetic resonance observation of Me_2Mg and $MeMgBr$ in a diethyl ether solution of methylmagnesium bromide. *Chem. Comm.* 1464 (1969).

The polarity of the ether and the degree of ionization of the MgX_2 component, are the primary factors (e.g., THF, dioxane and dimethoxyethane push the equilibrium to the right; and MgI_2 is very readily solvated to insoluble $[MgL_6]I_2$). In concentrated solutions, the composition is likely the X-bridged solvated dimer ($LRMgX_2MgRL$). And in very concentrated solutions both RMgX and R_2Mg form liner polymers with bridging halides and methyl groups.

Grignard reagents[100] are versatile in synthesis as a convenient source of nucleophilic carbon radicals (R-). In some cases, the Grignard reagent will act as a single-electron-transfer (SET) agent and react by this type of mechanism when the electrophilic center is easily reduced (as in the case of benzophenone) or not readily assessable to the R group and when the R group is relatively easily oxidized (as in the case of benzyl or t-butyl).

---

[100] V. Grignard. *Sur quelques nouvelles combinaisons organométalliques du magnèsium et leur application à des synthèses d'alcools et d'hydrocarbures* (On some new organometallic compounds of magnesium and their application to syntheses of alcohols and hydrocarbons). *Compt. Rend.* 130: 1322 (1900).

R-MgX + easily reduced electrophile (E) →
[R.MgX]$^+$ + E.$^-$

From this point, a variety of radical products are possible. In most of the intended reactions (e.g., with aliphatic aldehydes and ketones) the reaction is either nucleophilic or appears to be nucleophilic.[101]

8. Column Chromatography

In the late 1800s, petroleum exploration became a central issue with the US Geological Survey. Crude oil production doubled every decade from 1870-1900. Petroleum becomes trapped in domes under impervious layers of rock and soil. When you drill a well looking for oil, of course, you are hoping for early signs that you are on the right track. Thus, in 1897, David Talbot Day (1859-1915) of the USGS reported that petroleum produced different colored bands as it permeated upward into various types of soil. This idea was of substantial interest is drilling wells. But it was

---

[101] E.C. Ashby, J. Laemmle, H.M. Neumann. Mechanisms of Grignard reagent addition to ketones. *Account. Chem. Res.* 7(8):272-280 (1974).

probably not the origin of chromatography as we know it.

Mikhail Tsvet (1872–1919) was trained in physics but applied himself to botany in St. Petersburg, Russia. In the late 1890s, he was trying to isolate plant pigments and realized that some were more strongly absorbed to solids that other. By 1901 he presented a paper on separation of plant pigments by column chromatography using calcium carbonate as the stationary phase and mixtures of petroleum ether/ethanol as the mobile phase. The technique was first published in 1905 in Polish. By 1907, he had coined the term "chromatography" (color writing) and demonstrated the technique in Germany. Unfortunately, early attempts to reproduce the effects failed in the west and the idea largely fell into obscurity until it was re-discovered by Richard Kuhn (1900–1967) in 1929. Column chromatography, thus did not play a major role until what I call "early modern" organic chemistry (post-1930).

9. Studies in Natural Product Chemistry

There are many classes of natural products; here only a few of historical significance will be discussed.

Synthesis of Quinoline and its Derivatives

We have already mentioned the futile efforts of Perkin to synthesize quinine that led to his fortune in founding the synthetic dye industry (1856). Starting about 1880, there was a renewed interest in quinoline chemistry. Although some quinoline-based dyes had been formed as early as 1856, they were not commercially important until "quinoline yellow" was synthesized and made water soluble by sulfonation.

Quinoline Yellow

Drawing by Shaddack, source Wikimedia Commons

This event apparently contributed to a burst of interest in synthesis of quinoline derivatives from aniline:

Going clockwise from top: Combes quinoline synthesis (1888); Conrad–Limpach synthesis (1887); Doebner reaction (1887); Doebner–Miller reaction (1881-84); Gould–Jacobs reaction (1939); Skraup reaction (1880).

Full credit to Project Osprey for researching this chemistry, organizing this chart and providing the drawing and caption, source Wikimedia Commons.

The Skraup reaction (Zdenko Skraup (1850–1910))[102] is illustrative of the general approach. In a "one-pot"

---

[102]Z.H. Skraup. *Eine Synthese des Chinolins. Berichte* 13: 2086 (1880).

reaction (involving a series of steps) glycerol with sulfuric acid and a mild oxidizing agent (e.g., nitrobenzene) converts substituted anilines into the corresponding quinolines. The general steps are as follows:

(1) Acid-catalyzed eliminations of glycerol to acrolein (via the enol)

(2) 1,4-addition of aniline to the acrolein (Michael reaction)

(3) Internal Friedel-Crafts acylation to close the second ring

(4) Acid-catalyzed elimination and oxidation of the amine to form the quinoline

These reactions are particularly relevant to development of drugs in the 1930s.

The Quest for Synthetic Rubber

The demand for rubber skyrocketed in the 1890s. Based on the knowledge that isoprene was a building block of latex (un-vulcanized rubber) a considerable effort was made in an attempt to polymerize isoprene.

Some chemists reported success and others were not able to duplicate the results (1882 through 1913).[103]

It should be noted that the only source of isoprene that was known at the time (1913) was by very high temperature pyrolysis of turpentine. This proved to be a serious drawback even if isoprene could be polymerized for the following reasons: (i) pine sap was collected the same way that latex was collected (drop-by-drop) and its supply was limited; (ii) turpentine was distilled from pine sap (while latex did not have to be refined); (iii) the yield of isoprene from turpentine was low; and (iv) the quality of synthetic rubber in service was unproven. All of these factors, made commercialization of synthetic rubber via isoprene from turpentine improbable.[104]

In the course of attempting acid-catalyzed polymerization of isoprene in 1879, Georges

---

[103] F.J. Pond. A review of the pioneer work on the synthesis of rubber. *J. Amer. Chem. Soc.* 36 page 167 (1913). Overall, because of the economic importance of rubber after 1900, the literature circa 1910-1915 reflects a variety of claims and counter-claims driven by nationalistic, personal and financial motives.

[104] F.M. Perkin. Natural and synthetic rubber. *The Chemical Engineer.* XVIII(2):58-65 (1913).

Bouchardat appears to have achieved some degree of polymerization of isoprene, but mainly isolated mono- and di-hydrochlorides of isoprene (bp 86-91°C and 145-153°C, respectively), His "polymer" (C 87,1%, H 11.7% and Cl 1.7%) was not of high enough molecular weight to be considered rubber and retained some chlorine. By 1913, it had been found that the purest isoprene was obtained from the decomposition products of rubber by forming the dibromide, which could be purified and then reducing the dibromide with zinc. Prepared this way, isoprene boiled at 33.5°C.

In 1910, Sergei Vasiljevich Lebedev (1874–1934) made a useful polymer from butadiene and published his work in *Research in polymerisation of bi-ethylene hydrocarbons* (1913). His product found use during WWI; but after 1919, it was abandoned. He also developed a method for converting ethanol (from fermentation of potatoes) to butadiene by 1928; and commercialized a polymerization process for butadiene using metallic sodium. In the mid-1930s, this method was standardized in Russia. In 1927, Standard oil and I.G. Farben (Germany) signed agreements concerning development of synthetic gasoline and cracking products such as butadiene. In 1929, I.G. Farben succeeded in polymerizing isobutylene. However, through the 1920's, no one succeeded in making a "rubber" that competed with natural rubber in cost or utility.

Terpenes (Pinenes)

Some progress was made on the composition of turpentine and the structure of terpenes before 1870 (see above), but the structural basis of the relationship was not understood until the 1870s. The terpenes (isomers of $C_{10}H_{16}$) were generally divided between those derived from turpentine with boiling points around 160°C and those from orange peel with boiling points about 174°C.[105] William Tilden (1842–1926) introduced nitrosyl chloride as a reactant to form

---

[105] W.A. Tilden and W.A. Shenstone. XIX Isomeric nitrosoterpenes. *J. Chem. Soc.* 31:554 (1877).

Tilden's method of forming nitrosylchloride was very cumbersome and was replaced in 1888 by a method involving ethyl nitrite and HCl.

A Study of the Hydrolysis of Nitrosopinene to Hydroxylamine and Carvacrol By Joseph Alfred Hall, University of Wisconsin 1922. 0. Wallach, *Liebig Ann. Chem.* 246:221 (1888).

crystalline derivatives of terpenes in 1874[106] and began a systematic study (with Professor N. Story-Maskelyne) in 1877. Tilden found that those essential oils associates with turpentine formed a white, crystalline "nitroso chloride compound" ($C_{10}H_{16}NOCl$) that yielded a single compound $C_{10}H_{15}NO$ after treatment with sodium hydroxide. The use of excess acid especially at ambient temperature or above produced tarry products.

In 1893, Julius Bredt (1855–1937) determined the structure of camphor and G. Wagner deduced the structure of pinene the next year (1894).

Camphor   **Alpha- pinene**

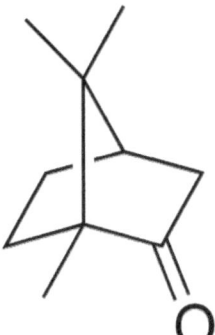
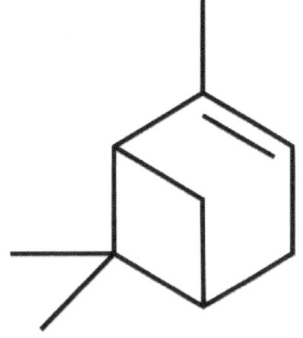

Drawing by Edgar181,   Drawing by J.delanoy, source
source Wikimedia Commons   Wikimedia Commons

---

[106] W.A. Tilden. XL.On the action of nitrosyl chloride on organic bodies. – Part I. On phenol. *J. Chem. Soc.* (London), 27:851-852 (1874).

As work was done on these compounds, it became apparent that under acid conditions, they underwent various rearrangements. These became known as the Wagner-Merrwein rearrangements (i.e., shifts of H or R groups) to form the most stable carbocation after the work of Hans Meerwein (1879–1965).[107]

Nuclein and Nucleobases

Physician Friedrich Miescher (1844–1895) isolated nuclein (DNA and associated proteins, e.g., histones) from the nuclei of leucocytes in 1869. In 1878, Albrecht Kossel (1853–1927) separated nucleic acid from the histones and succeeded in isolating and naming the five phosphate-containing nucleobases between 1885 and 1901: adenine, cytosine, guanine, thymine, and uracil. Concurrently, he isolated histidine and worked out methods to separate the hexose bases (alpha-amino acids arginine, histidine, and lysine). In 1884, Adolf Pinner (1842–1909) condensed ethyl acetoacetate with amidines and called his product "pyrimidine" in 1885.

---

[107] G. Wagner. *J. Russ. Phys. Chem. Soc.* 31: 690 (1899).
H. Meerwein. *Über den Reaktionsmechanismus der Umwandlung von Borneol in Camphen. Justus Liebig's Annalen der Chemie* 405(2): 129–175 (1914).

Purine

Drawing by NEUROtiker, source

Wikimedia Commons

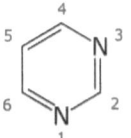

Pyrimidine

Drawing by Jynto, source
Wikimedia Commons

In 1889, Emil Fischer synthesized purine from uric acid. In 1900 Wilhelm Traube (1866–1942) found that he could make purines from pyrimidines by condensing an amine-substituted pyrimidine and formic acid. [108] He also succeeded in the total synthesis of caffeine from N, N'-dimethyllurea and ethylcyanoacetate.

Caffeine

Author: NEUROtiker; Source Wikipedia

---

[108] It has been discovered that formamide condenses to purine (170 °C for 28 hours).

Between 1909 and 1929, Phoebus Levene (1863–1940) greatly clarified the nature of DNA and RNA. He identified ribose and deoxyribose sugars and the nucleotides (phosphate, sugar-base) groups and realized that DNA was a combination of the various nucleotides held together by phosphate ester bridges. However, he did not anticipate that the polymer (which he though was a repeating tetrad) could convey information. At the time, proteins were believed to be the source of inheritable information.

Meanwhile, biologists discovered chromosomes in the nuclei of cells. Moreover, by 1905, Walter Sutton (1877–1916) and Theodor Boveri (1862–1915) stated that chromosomes must be the basis of inheritance. The work of Gregor Mendel on inheritance was rediscovered and Thomas Hunt Morgan (1866-1945) led his research group looking for mutations in fruit flies. Their work was conclusive that chromosomes carried genes (1915). Indeed, by 1923, Theophilus Painter (1889–1969) had counted the human chromosomes (karyotype).

10. Roots of the Pharmaceutical Industry

Until the work of Pasture in the 1870, little was known about the nature of disease. The idea that many human diseases were caused by microorganisms sparked research to find cures.

Salicylate and Aspirin

Through the American Civil War (1861-65) gangrene turned many survivable wounds of the arms and legs into death sentences.[109] At the time, the nature of infection was not understood. The connection between microorganisms and disease was only slowly growing in the 1860s. Nonetheless, in part from Louis Pasture's demonstration that beer and wine did not go bad

---

[109] Confederate General Thomas Jackson was the victim of friendly fire and suffered several wounds from which he appeared to be recovering before he died suddenly apparently of blood poisoning. The author's mother's grandfather sustained a relatively minor wound on the lower arm and subsequently had multiple amputations of the arm ending in the upper arm as the surgeons attempted to stay ahead of the infection.

without exposure to sources of infection, the idea that it might be able to prevent or overcome infections by chemical means gained some ground. In 1867, Joseph Lister (1827–1912) started applying carbolic acid (phenol) to wounds to prevent infection and as a method of sterilizing surgical instruments. His success attracted substantial attention by 1873. By this time, the corrosive effects of phenol on skin was recognized and the idea came to Herman Kolbe that the compound that he and Schmitt had synthesized in 1863 (salicylic acid), might be less objectionable than phenol while retaining the anti-microbial effects. He did some experiments with salicylate and successfully prevented fermentations. He also found that, salicylate could be tolerated internally and that it seemed to have antibiotic effects because it reduced fever and pain.

Meanwhile, Friedrich von Heyden (1838-1926) received his doctorate as a student of Schmitt in 1874. Von Heyden set up a factory to manufacture salicylate. Their product proved to be very popular for pain and fever, although it was found not to be an effective antibiotic. In 1885, Schmitt's most successful student Richard Seifert joined the factory.

In a different venue, Felix Hoffman (1868-1946), working for the Bayer[110] Company, synthesized acetylsalicylate (ASA) in 1897. In this process, it was clarified that the acetyl group was on the phenol (forming an ester); i.e., not on the carboxylate forming an anhydride (and not on the aromatic ring, see Friedel-Crafts acylation). The motivation for the work was to find a less irritating material than salicylate from the von Heyden Company. ASA is much better tolerated than salicylate itself[111]; and under the trade name Asprin®, they sold their product around the world by 1899. In 1901, the von Heyden Company also began selling ASA in Germany (where there was no patent) and in other countries where Bayer had obtained patents. Obviously, litigation followed and it was resolved that the Bayer patents had been incorrectly issued because the work of Gerhardt, Kolbe and others (before 1870) had been overlooked.[112]

---

[110] Be careful not to confuse Bayer (i.e., Friedrich Bayer (1825–1880)) with Baeyer (Adolf von Baeyer [pronounced "Buyer" in English] (1835–1917)).

[111] Presumably because of the slow hydrolysis of the ester releasing the phenol.

[112] The reader should note that some of the popular literature has been slanted to favor one party or the other.

It is worth mentioning that during WWI (1914-18) production of ASA was limited by the availability of phenol. Thomas Edison had set up a phenol plant to make phenol for Bakelite phonograph records, and while the U.S. was still neutral in the war, Germany arranged to buy all of Edison's excess phenol (at a very high price) and route it to their salicylate-manufacturing subsidiary in the U.S.; from there the salicylic acid was shipped to Germany. This was all legal, but because the German and British phenol shortage was caused by the use of phenol to make picric acid (trinitrophenol)[113] explosive the direct shipment of phenol to Germany would have raised eyebrows. This became known as the "phenol plot;" and when it was made public, Edison sent his excess phenol to the U.S. military. When the U.S. entered WWI in 1917, German property in the U.S. became subject to the "alien property custodian" who was responsible to see that it was not used to support the

---

[113] Picric acid can be prepared directly from benzene by the Wolffenstein–Böters reaction. The reaction is said to involve formation of the phenol (via the diazonium) by the actions of mercury nitrate and nitric acid. R. Wolffenstein and O. Böters . *Über die katalytische Wirkung des Quecksilbers bei Nitrierungen. Berichte. der deutschen chemischen Gesellschaft* 46 (1): 586–589 (1913).

enemy nation. Ultimately, Bayer's company (including the intellectual property, trademarks, brand names, etc.) was seized and auctioned to an American company. The valuable marketing tools were ultimately recovered by Bayer for approximately a billion dollars in 1994.

Toxicology and Pharmacology (Ehrlich)

In the 1870s, the association of microorganisms with disease had greatly increased interest in microscopy and identification of bacteria and protists (protozoa). Karl Weigert (1845–1904) was one of the first to realize that bacteria absorbed chemical dyes and passed this research idea to his cousin Paul Ehrlich (1854–1915). Ehrlich was also studying medicine, but his interest in dyes caused him to develop a substantial background in chemistry so that he understood the general composition, solubility and acid-base characteristics of dyes. With this knowledge, Ehrlich conducted groundbreaking research on blood cells. His dissertation (1878) was entitled "Contributions to the Theory and Practice of Histological Staining" and he managed to identify and name a number of different types of blood cells by the way they stained with various dyes. He was at an intersection of chemistry and biology that at that time had no clear definition and, hence, his career followed several paths.

He soon (1882) became a friend of Robert Koch (1843-1910). By 1885, Ehrlich was applying staining *in vivo*. This work showed that a chemical could be introduced to the body of a live animal and go to selected targets. Work with Koch led him into immunology and the development of serums against diphtheria, tetanus and other diseases. In these diseases, the adverse effects on humans are the result of toxins release by the organism (not merely the multiplication of the organism). A "serum" is the cell-free and clotting–factor-free blood plasma from an animal that has been progressively exposed to a toxin (e.g., produced by a micro-organism) to the point that it has developed a level of immunity (i.e., levels of anti-toxin) that neutralize the toxin. Ehrlich developed techniques to quantitatively evaluate the potency of toxin and anti-toxin solutions. It was clear to Ehrlich that he was dealing with chemical reactions (e.g., with definitive stoichiometry).

After finding that methylene blue dye is concentrated in malaria organisms and other *Plasmodiidae*. Ehrlich injected methylene blue dye into patients with malaria and discovered that it seemed to kill the microorganism; but it had adverse effects on the patients (1889). In experiments with laboratory animals, he also found that the dye *trypan red* would kill protozoa that cause African sleeping sickness; hence, they were called *trypanosomes*.

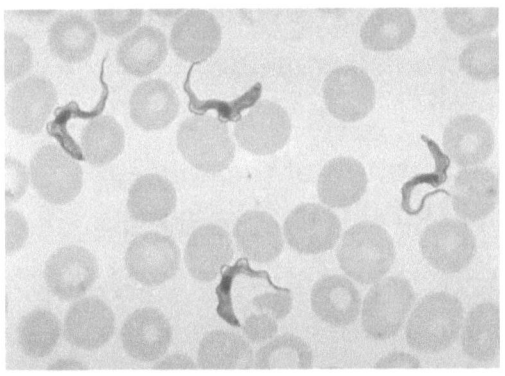

Trapanosomes stained with Trypan Red

Image from the US Centers for Disease Control

Dr. Myron G. Schultz, source Wikimedia Commons

Through this work, Ehrlich developed the idea (1897) that organisms have "side-chains" (e.g., surface proteins) that are necessary for their normal function and that toxic agents can combine with the sidechains causing them to malfunction. Anti-toxins, thus, may mimic the side-chain to trap and neutralize the toxin. Concepts of toxicology and pharmacology predate Ehrlich, but he made the ideas into a quantitative science.

Arsenic (As_2O_3) had been recognized as a bioactive agent in antiquity and in the 1700s a 1% solution of potassium arsenite (Fowler's solution) had been introduced as a general tonic and it was specifically used against the symptoms now recognized as leukemia (1845).[114] However, the liberal use of Fowler's solution also was anecdotally associated with skin cancer (1887, J. Hutchinson) and this has been taken as an assumption of *cause and effect* ever since.[115] Nonetheless, in the early 1900s the association of arsenic with cancer was still received skeptically. A much less toxic form of biologically-active arsenic (arsenic acid, "Atoxyl") was discovered in 1859 by Antoine Béchamp (1816–1908). And, this compound of

---

[114] Arsenic compounds have recently been recognized as having a favorable therapeutic effect in various leukemias: X. Huang et al. Arsenic trioxide induces apoptosis of myeloid leukemia cells by activation of caspases. *Med Oncol.* 16(1):58-64 (1999).

[115] In particular, the appearance of skin cancers in SW Taiwan in the 1950s was assumed to be due to arsenic in drinking water (there are many flaws in this study) and this has been the basis for the strict standards placed on arsenic in drinking water by the U.S. EPA. I address this issue in my book *The Myth of the Linear, No-Threshold Dose-Response for Carcinogens* (2013) published through Amazon Kindle.

arsenic was being evaluated for use as a drug *circa* 1900.

Atoxyl

Drawing by Smokefoot, source Wikimedia Commons

African sleeping sickness was one of the major barriers to European exploitation of Africa. Thus, colonial countries were studying it intensely in the late 1800s. It was not until 1903 that African sleeping sickness was realized to be the extreme result of a mild fever caused by trypanosomes in the blood transmitted by biting flies.   In 1905, it was discovered that Atoxyl was effective against some species of trypanosomes. Koch and others started using Atoxyl in patients. The drug was found to be effective in the early infection but not affective in late infection (where the parasite had crossed the blood-brain barrier); and it was also found that high doses of Atoxyl were especially toxic to the optic nerve causing blindness.

Based on these observations, Ehrlich had his organic chemist, Sahachirō Hata (1873–1938), make arsenic derivatives, which he then tested for efficacy against various organisms and toxicity to humans. They discovered that arsphenamine (Salvarsan, compound 606) was effective against syphilis and tolerated by humans. The empirical formula of this compound led them to formulate it as an analogue of azobenzene, but it has been found recently to actually have a five-membered ring of arsenic atoms (As-As bonds). Ehrlich's success (1909) brought him wide fame and the term "Magic Bullet" was coined to describe the idea of a targeted antimicrobial that was safe and effective.

The 1906 Pure Food and Drugs Act

The first government agency in the U.S. to practice chemistry was the Patent Office where Lewis Beck began evaluating claims for agricultural products in 1848. This responsibility passed to the Bureau of Chemistry of the Department of Agriculture in 1862.

Harvey Wiley (1844–1930) was humbly born in rural Indiana, was an enlisted man in the Union Army (1864-65) and was educated through a series of steps leading to a B.S. degree in chemistry from Harvard in 1873. He returned to Indiana on the faculty of Purdue University (1874) and in 1878 he went on sabbatical in Germany where he studied under Hofmann and learned to use a polariscopic to analyze optically active sugars. Back in Indiana, he began checking sugar for adulteration with glucose on behalf of the state government (paper published 1881). His reputation for objectivity and his skill at analysis of sugar spread and he was offered the job of Chief Chemist of the USDA Bureau of Chemistry in 1882. After settling issues concerned with the economics, safety and politics of sugar, broader issues caught his attention.

The American Chemical Society had been formed April 6, 1876; and Wiley became its president (1893-94).

In 1896, New Jersey housewife Alice Lakey (1857–1935) became interested in the purity of foods (apparently because her father was described as a "picky eater"). After ascending to the top of a local homemakers; association, she invited Harvey Wiley to speak. They soon formed an alliance of common interests in ensuring the purity of food and drugs. Back in Washington, Wiley began to lobby Congress for legislation to regulate interstate commerce in food, while Lakey spread the word among millions of housewives who were also becoming motivated by the suffrage movement. When reform-minded Theodore Roosevelt became president of the U.S. as a result of the assignation of President McKinley, the gate was opened for food and drug regulation. In 1902, Wiley created his "Poison Squad" of young men eating foods with various additives, which (of course) received much media attention. The final motivating impact was provided by Upton Sinclair, Jr. (1878–1968) who worked briefly in the meat packing industry and released a serialized novel entitled *The Jungle* exposing abuses and working conditions in 1905. With a million letters from housewives and the mid-term elections scheduled for November, the Pure Food and Drug Act was passed and signed by the President in 1906.[116] The administration of this act was handed to Wiley in the Bureau of Foods.

Arsenical, Sulfur and Quinoline Drugs

We have already mentioned Arsphenamine (compound 606) and Atoxyl. The discovery of these compounds created great enthusiasm among doctors and chemists. At the Rockefeller Institute in New York, chemists Walter Jacobs and Michael Heidelberger (1888-1991) began synthesizing arsenic compounds while Wade Hampton Brown and Louise Pearce (1885-1959) evaluated them in animal bioassays.

In 1919, Heidelberger and Pearce were working on "compound A63" (the 63rd variation on Atoxyl) and believed they had come up with something that

---

[116] This is a textbook strategic plan for getting legislation passed in the U.S. Congress. We will see a similar pattern for the establishment of the U.S. Environmental protection Agency in 1970.

worked very well against early *and* late stage sleeping sickness.

Tryparsamide (A63)

Drawing by Yikrazuul, source Wikimedia Commons

In 1920, Pearce travelled to Leopoldville (Kinshasa), in the Belgian Congo, where there was a Belgian clinic. The clinic received sleeping sickness patients from Cameroon via the riverboat trade on the Congo and its tributaries (i.e., the Sangha River). The trials were successful and A63, became known as Tryparsamide and was regarded as the first effective drug against sleeping sickness.

11. Reduction of Aldehydes and Ketones to Alcohols

Before introduction of metal hydrides[117], several hydride transfer reactions were discovered.

Meerwein-Ponndorf-Verley (MPV) Reduction

In 1851, Stanislao Cannizzaro (1826–1910) took time out from trying to rationalize the system of atomic masses consistent with the ideas of his countryman Amedeo Avogadro (1776–1856) and discovered that aromatic aldehydes disproportionate in the presence of base to yield equal molar amounts of alcohols and carboxylic acids. The Meerwein-Ponndorf-Verley (MPV) reduction appears to share some mechanistic features.

$$\text{Al(i-Pro)}_3 \text{ / heat}$$

OH   O   Oppenauer oxidation   O   OH

R_1 R_2 H_3C CH_3 Meerwein-Ponndorf-Verley reduction R_1 R_2 H_3C CH_3

Drawing by Lankaluf, source Wikimedia Commons

---

[117] H.C. Brown and S. Krishnamurthy. Forty years of hydride reductions. *Tetrahedron*. 35:567-607 (1978).

P a g e 178 | 483

In 1925-26, three groups independently recognized that aluminum alcoholates and ketones engage in an equilibrium to produce the most stable combination of products. By using an excess of isopropanol, the equilibrium can be pushed towards reduction of most ketones. The groups that made these discoveries were headed by Hans Meerwein[118] (who was known for his carbocation work), Valery[119] and Ponndorf.[120] The reverse (oxidation) reaction is named after Rupert Oppenauer (1910-1969).

Wolff-Kischner Reduction

Reactions of hydrazine or its derivatives (e.g., semicarbazene) with a ketone forms a hydrazone (or semicarbazone). Nikolai Kischner (1867–1935) and Ludwig Wolff (1857–1919) independently published

---

[118] H. Meerwein, R. Schmidt. *Ein neues Verfahren zur Reduktion von Aldehyden und Ketonen. Justus Liebigs Annalen der Chemie.* 444(1):221–238 (1925).

[119] A. Verley. *Bull. Soc. Chim. Fr.* 37: 537 (1925).

[120] W. Ponndorf. *Der reversible Austausch der Oxydationsstufen zwischen Aldehyden oder Ketonen einerseits und primären oder sekundären Alkoholen anderseits. Angewandte Chemie.* 39(5):138–143 (1926).

papers in 1911 and 1912, respectively, describing the reduction of ketones via their hydrazones:

Drawing by C kleinlein, source Wikimedia Commons

This approach (which is not compatible with base-sensitive substrates) has been modified in various way to adapt to specific synthetic schemes.

Clemmensen Reduction

Erik Clemmensen (1876-1941) was born in Denmark and immigrated to the U.S. in 1900. In 1913, he invented a method of reducing ketones using zinc (mercury amalgam) with hydrochloric acid. Thus, his reaction conditions complement the Wolff-Kischner conditions.

12. The Automobile and Gasoline (1908-1930)

Henry Ford was an engineer with Edison's company; but by the late 1800s, he managed to found a small automobile company. After an amazing story of inventiveness (which included designing and building an engine, investing in a steel mill to produce high-strength vanadium steel, and re-thinking the manufacturing process) he introduced a reliable car (the Model T) that was affordable by the general public in 1908. Increased public interest in personal transportation produced a need for paved roads and a different type of liquid fuel. Most of the early internal combustion engines had run on the gases isolated from distillation of kerosene, but it was not feasible to carry or store "gas" for transportation. Fortunately, the float-type carburetor had been developed by Wilhelm Maybach and was widely available by 1900. Thus, with the introduction of electric lights (by Edison and Westinghouse) and the automobile by Ford, petroleum refining turned from kerosene to "gasoline" a more volatile cut (with higher purity requirements) and asphalt became an important byproduct.

Of course, optimizing the production of gasoline from a barrel of crude oil became an economic driver for the petroleum industry as demand increased and crude oil production "stabilized" (relative to what it had been) before the discoveries of the east Texas oil fields (1930-1934).

Another factor also entered the calculation: octane rating. In the 1920s, internal combustion engine designers realized that greater thermal efficiency required higher compression engines. The problem of pre-detonation was encountered and the octane scale was introduced.

Straight Run Gasoline

Natural petroleum (crude oil) is a complex liquid. It has few direct uses, but it is a raw material that can be refined and transformed into many useful products. The original refining process utilized until about 1930 was nothing more than fractional distillation. Over the years, the preferred fraction of petroleum changed. In the 1800s, petroleum was typically fractionated to obtain kerosene for lamps and lubricating oils and greases for steam-driven machinery. At this time the low boiling liquids, called "naphtha," had few uses and were either discarded or used as solvents.

The original internal combustion engines used the volatile gases from petroleum (butane and pentane). But these gases were in relative short supply and difficult to store. The inventors discovered that flammable mixtures of naphtha could be prepared with air and soon they were tinkering with a device called a "carburetor." The carburetor was designed to vaporize the naphtha into a moving stream of air at just the right rate to form a combustible mixture of hydrocarbon (1.2% to 7.1% V:V) with air. The fraction of naphtha that worked best for this process was called "gasoline."

The early engines had low compression ratios and used very low octane fuel (*e.g.*, the naphtha fraction of petroleum has an octane of around 50). Engineers soon realized that the horsepower-to-weight ratio of engines could be increased by using higher compression engines. It was not long before the phenomenon of "knock" was discovered, and the need for higher octane fuel was created. Starting in 1923, tetraethyl lead (TEL) was added to gasoline and this was able to increase the octane by as much as 30 octane numbers. Through the 1920s, straight run[121] gasoline with "lead"

---

[121] The modifier "straight run" was added later to distinguish gasoline obtained from the continuous distillation run from gasoline obtained by blending or chemical modification. Since WWII, gasoline has been *synthesized* from petroleum rather than *refined* from it. The petroleum companies have retained the term refining in part to disguise their relationship to the rest of the synthetic chemical industry especially once the environmental movement began. Even as late as 1998, the U.S. Environmental Protection Agency labored under the assumption that gasoline is a natural fraction *refined* from petroleum.

(TEL) was effective. The following table summarizes the Brake Thermal Efficiency as a function of the compression ratio. The corresponding required octane rating of the fuel is identified.

Brake Thermal Efficiency (full throttle)	Compression Ratio Required	Fuel Octane Required to Achieve Compression Ratio
low	5 to 1	72
25%	6 to 1	81
28%	7 to 1	87
30%	8 to 1	92
32%	9 to 1	96
33%	10 to 1	100

Tetraethyl-lead

Shortly, after the First World War, General Motors Corporation began to make gasoline engines more efficient. Soon, the favorable effect of increased compression ratios was known and the phenomenon of engine "knock" was discovered. Naturally, there were consultations with the major oil companies and an arbitrary scale of resistance of fuel to "knock" was developed using isooctane (i.e., 2,2,4-trimethylpentane, which had favorable anti-knock properties) as the standard. Experiments to find agents to maximize the octane number were soon underway with the collaboration of chemists and mechanical engineers. Edisonian experimentation led to the discovery of the favorable effects of tetraethyl-lead. When added to gasolines of the 1920s, the octane ratings could be raised by as much as 30 octane numbers. However, there were economic and technical limits to this effect, and for practical purposes the increase in the maximum achievable octane was about 10 octane numbers (*e.g.*, from about 80 to about 90 octane).

General Motors and Standard Oil teamed up to form the Ethyl Gasoline Corporation to manufacture and sell tetraethyl-lead. The synthesis involved reaction of ethyl chloride with sodium-lead amalgam:

$$4\ Na(Pb) + 4\ C_2H_5\text{-}Cl \longrightarrow (C_2H_5)_4Pb + 4\ NaCl$$

Tetraethyl-lead (TEL) was isolated by steam distillation and was found to be a liquid that boiled at 200°C with a density of 1.62 grams *per* mL. Within a year of its introduction in February of 1923, the toxicity of TEL was recognized. The additive was removed from the market for a year; and when it was reintroduced in 1925, the concentration of TEL in gasoline was limited to 3.0 mL *per* gallon. When TEL was first introduced in 1923-25 its price was about $3.00 *per* pound. This dropped rapidly to about $0.75 *per* pound by 1935 and stabilized at about $0.50 *per* pound during the war years (1939-45).

It was soon recognized that TEL produced lead deposits in the combustion chamber when it was used by itself. The TEL molecule was destroyed during combustion and lead and lead oxides were not volatile. To counter this feature, halogenated additives were included in the "Ethyl Fluid" package:

TEL	63.30%
Ethylene Dibromide	25.74%
Ethylene Dichloride	8.73%
Dye and impurity	2.23%

The dye was included to warn users that the product contained toxic TEL. Blue dye was specified in 100-

octane U.S. Army grade "Fighting 2-92" and U.S. Navy grade "M 302."

Typical Products of Fractional Distillation of Crude Oil			
Boiling Range (degrees Celsius)	% Recovered by Atmospheric Distillation	Chemical Composition	Uses
0-30	2%	C1-C4 (butane/ butene)	natural gas fuel and feedstock
30-200	15-30%	C5-C12	"straight run" gasoline (about 60 octane) and naphtha
200-300	5-20%	C12-C15	kerosene/jet fuel
300-400	10-40%	C15-C25	gas oil/ bunker fuels for ships
over 400	10-50%	over C25	Residue/ cracking stock/asphalt

13. Niacin and Ascorbic Acid

The Concept of Vitamins

Following the discovery that many diseases of humans and animals are caused by microbes by the chemist Louis Pasture (circa 1865), microbiologists worked diligently to discover the cause of various diseases. But some diseased did not follow a normal pattern of contagion and did not present infective agents. These diseases included scurvy and beriberi, both of which were known to be most common among seamen. Scurvy became known in Europe during the Crusades (circa 1300) and it became the major factor limiting ocean exploration in the 1500s. By the 1700s, it was known that fruits of citrus plants provided cures and the British, Spanish, French and Portuguese all planted citrus trees along sea routes to mitigate the problem.

Beriberi was discovered by the Dutch in the East Indies (Java) in the 1600s, where rice was the main source of calories. But it was not until the late 1800s when the Japanese began long-distance seaborne travel (e.g., 1885) that beriberi became an important disease. While working in Indonesia, Christiaan Eijkman (1858–1930) noticed that chickens fed military standard polished rice for months developed beriberi but recovered when returned to ordinary unpolished rice. Thus, he deduced that the husk of the rice contained a nutrient that prevented beriberi. In Europe, the Norwegian fishermen working on the Grand Banks off Nova Scotia and whaling around the world were also developing beriberi.

Axel Holst and Theodor Frølich followed Eijkman's discoveries and applied the same dietary scheme that Eijkman took to guinea pigs, which coincidentally share the inability to make ascorbic acid found in humans.[122] By 1907, they found that the limited diet produced scurvy in guinea pigs, which they were able to cure with fresh vegetables.[123]

Meanwhile, Casimir Funk (1884–1967) received an education, studied trace elements in humans and began a study of amino acids (cysteine and alanine) with Emil Fischer in 1904. In 1910, Funk moved to the Lister

---

[122] Around 60 million years ago, the primate line sustained a mutation in its L-gulono-γ-lactone oxidase (GLO) gene, which only affected the production of ascorbic acid. Guinea pigs sustained a different mutation about 14 million years ago. M.Y. Lachapelle and G. Drouin. Inactivation dates of the human and guinea pig vitamin C genes. *Genetica*. 139(2):199-207 (2011).

[123] K.J. Carpenter. The discovery of vitamin C. *Ann Nutr Metab*. 61(3):259-64 (2012).

Institute in London where he became interested in beriberi. He realized that the husk of rice contained an essential nutrient and soon isolated a fraction that contained the nutrient (i.e., niacin). Here he coined the word "vita-amine" (live giving-amine), which was shortened to vitamin when it was realized that not all vitamins contain nitrogen. In 1912, he proposed that each of the diet-related diseases was caused by lack of one or more vitamins.

Isolation and Synthesis of Ascorbic Acid

Like many others Albert Szent-Györgyi (1893 –1986) was swept into military service in 1914. He left service in 1916 as a result of a wound, which he later admitted was self-inflicted. He received his doctorate in 1917 and then did post-doctoral work at several laboratories in Europe. By 1922, he was particularly interested in cellular metabolism and energy production. He isolated a compound from adrenal glands that acted as a hydrogen carrier gaining and losing hydrogen atoms in the course of cellular metabolism. The compound had six carbon atoms and behaved as a carboxylic acid ($C_6H_8O_6$) so he called it "hexuronic acid." But he only had traces of it at that time. In 1930 with J. L. Svirbely, he found that hexuronic acid prevented scurvy in guinea pigs. Isolation of large quantities of hexuronic acid proved difficult because most sources (citrus juices) also contained sugars that were hard to separate.

Finally, Szent-Györgyi discovered that red paprika also contained large quantities of hexuronic; and in 1933, he isolated kilogram quantities.

L- Ascorbic Acid

Drawing by Yikrazuul, source Wikimedia Commons

He sent samples to Norman Haworth (1883-1950) at the University of Birmingham and other research groups. A race followed to identify and synthesize the natural product. Tadeusz Reichstein's (1897–1996) research group in Switzerland synthesized D-ascorbic acid from D-xylosone, which was readily available in 1933. But it was clear that they had the wrong optical isomer.

The Haworth team deduced the correct structure and successfully synthesized L-ascorbic acid in 1934. Their approach required the L-isomer of xylosone, which they isolated with great difficulty through the Fisher's osazone process (i.e., formation of phenylhydrazine derivatives and separation of diastereomers by crystallization) starting with D-galactose.

In the meantime, Reichstein's research group developed a more practical synthesis starting with D-(+)-glucose. The Reichstein process is still the basis of commercial synthesis of vitamin C. After reducing D-glucose to D-sorbitol, the key innovation is use of *acetobacter* to provide the correct stereochemical oxidation of sorbitol to L-sorbose. In the process patented in 1935, a protective acetal was formed before oxidation with permanganate. However, in 1942, Kurt Heyns (1908–2005) discovered that the oxidation can be done in one step with oxygen over platinum catalyst. The 2-keto-L-gulonic acid forms a lactone with elimination of water. Then the ketone tautomerizes to the enol.

14. X-Ray Crystallography: Direct Proof of Structure

An important byproduct of the discovery of x-rays by Wilhelm Röntgen (1895) was their application to explore the nature of crystals. By 1915 the father and son team of W. H. Bragg and W. L. Bragg had worked out the principal problems in determining distances between planes of atoms that reflected x-rays of specific length. The methodology was published in a book (*X-Rays and Crystal Structure* (1915)) by W. L. Bragg and it includes calculations of the distances between planes and density of the crystals. These results immediately suggested that ions of salts were not selectively clumped together as molecules, but rather formed an alternating lattice. Their work, including the structures of diamond and graphite, soon followed. These structures reflected on the structure of cyclohexane and benzene.

At that time, there was much unresolved speculation about the benzene ring. In 1926, Linus Pauling used the "electron-orbit-sharing theory of the chemical bond" to predict that it would be planar. For organic chemists, Kathleen Lonsdale (1903 - 1971) was the first to determine the structure of a benzene-ring (1928, specifically hexamethylbenzene). Her work[124] confirmed the ring structure, and showed that the ring was planar, but surprised chemists by demonstrating that all the C-C bonds were the same length (not alternating double and single bonds). She determined the interior bond angles to be 119°34′ and C-C distances

---

[124] K. Y. Lonsdale. The Structure of the Benzene Ring. *Nature* 122(810): 122 (1928).

of 0.142-0.148 nm. In 1931, she determined the structure of hexachlorobenzene.

15. Organic Reaction Mechanisms

With acceptance of the structural hypothesis, it became apparent to organic chemists that reactions must involve some sort of organized rearrangement of atoms from the reactants to the products. Clearly, bonds were being broken and bonds were being made. But, how and why?

Kinetics

Several interesting kinetic studies had been reported in the literature. Starting in 1850 with Wilhelmy's report of the hydrolysis sucrose into D-(+)-glucose and D-(-)-fructose catalyzed by acid. Thus, it is not surprising that van't Hoff who had recently (1874) championed the structural organization around carbon atoms would be the first to think about the dynamics of organic chemical reactions. Fortunately, he was a fairly good mathematician and developed differential equations for forward and reverse reaction and to explain equilibria and introduced terms to account for the effects of temperature on reaction rates. This was published in 1884 as "Studies of Chemical Dynamics" (*Études de dynamique chimique*), a revised and enlarged translation was published in 1896.[125] In the meantime, Ostwald published a textbook *"Lehrbuch der allgemeinen Chemie"* in 1887. By 1896, van't Hoff had described first order, second order and higher order reactions; gas phase, liquid phase and heterogeneous reactions; and considered the effects of temperature on chemical reactions. On page 20, of his book (1896 translation) he uses the term "mechanism." In the text, it is clear that van't Hoff understood that the rate was determined by certain elementary steps and not the overall reaction stoichiometry. In 1889, Svante Arrhenius (1859-1927) followed van't Hoff and developed a more theoretically based rationale for the temperature dependence of reaction rate,

$$k = A \exp(-B/T)$$

and interpreted it as an "energy barrier" requiring an "activation energy." Thus, by 1890, chemists were in a

---

[125] This book is available on line at
http://babel.hathitrust.org/cgi/pt?id=uc2.ark:/13960/t4pk08n9
4;view=1up;seq=25

position to discuss chemical reaction mechanisms and explain chemical reactions.

Intermediate Carbocations

In order to explain some transformations, it seemed that there must be relatively unstable intermediates with unusual valences. In 1891, Merling had reported that tropylidene (cycloheptatriene) reacted with bromine to produce a crystalline water soluble product.[126] Although the structure was unknown, it appeared that the organic radical carried a positive charge.[127] In 1899, Julius Stieglitz (1867–1937)[128] suggested the possibility of unstable carbocations as reaction intermediates and these were soon (1901) demonstrated with triphenylmethyl cations by J. F. Norris and others.

---

[126] G. Merling. Ueber Tropin. *Berichte der deutschen chemischen Gesellschaft*. 24:3108–3126 (1891).

[127] The stability was not explained until the work of Erich Hückel (1931) and the structure was confirmed by Eggers Doering and Knox (1954).

[128] J. Steiglitz. J. *Am. Chem. Soc.* 21, 101 (1899).

Intermediate Radicals

Moses Gomberg (1866–1947) demonstrated the presences of the triphenylmethyl radical in 1900. He had successfully prepared tetraphenylmethane in 1898 by reacting triphenlymethyl chloride with phenylhydrazine and oxidizing the triphenylmethyl phenylhydrazine to the corresponding azo compound, which split out nitrogen (N_2) at 120°C. Following this success, he tried to prepare hexaphenylethane by the Wurtz coupling reaction. But what he isolated proved to be the peroxide R-O-O-R.[129] In the absence of oxygen (under CO_2) he reacted zinc with triphenylmethyl chloride and obtained the free radical, which is in equilibrium with a coupling product.[130]

---

[129] The stability of the peroxide suggest that steric effects are at play here, but resonance stabilization of the radical (structure 2) explains why it is in equilibrium with structure 3. L. Pauling and G.W. Wheland. The nature of the chemical bond. V. The quantum-mechanical calculation of the resonance energy of benzene and naphthalene and the hydrocarbon free radicals. *J. Chem. Phys.* 1: 362-374 (1933).

[130] M. Gomberg. *Chem. Ber.* 33, 3150 (1900). *J. Am. Chem. Soc.* 22, 757 (1900).

Scheme for the synthesis of trivalent carbon by Moses Gomberg

Full credit for drawing and caption by DrBurningBunny, source Wikimedia Commons

Intermediate Anionic Carbon

A number of complex reactions involving anionic carbons had been observed.

Mechanism of the Benzoin condensation

Full credit for drawing and caption by Brianlee89, source
Wikimedia Commons

The condensation of benzaldehyde to benzoin in the presence of cyanide had been known since the work of

Wöhler and Liebig (1832) and had been efficiently
carried out by Nikolay Zinin in the late 1830s. But how
it occurred was a mystery until Arthur Lapworth
(1872–1941) proposed the involvement of an anionic
intermediate in 1903.Lapworth is credited with being
one of the first chemists to think in terms of
mechanisms of reactions. His idea was that within a
molecule alternating centers of positive and negative
charge were induced during chemical reactions that
these polarities determined the course of a chemical
reaction.

Complex Rearrancements

In 1904, Holleman and J. Potter van Loon became the
first (of many)[131] to study the kinetics and mechanism
of the rearrangement of hydrazobenzene to benzidine.

Drawing by Edgar181, source Wikimedia Commons

---

[131] H. J. Shine. A personal history of the benzidine
rearrangement. *Bull. Hist. Chem.* 19: 77-92 (1996).

They found that the reaction followed a rate equation with a second order dependence on acid (protons)[132]:

$$Rate = d[benzidine]/dt = k[HCl]^2$$

The transformation is due to the hydrogen ions of the acid, for on comparing the action of hydrochloric acid and dichloroacetic acid the reaction constant was shown to be proportional to the degree of ionisation of the acids employed. This caused Dr. van Loon to suggest that during the transformation two H-ions are first linked to hydrazobenzene forming

$$C_6H_{5\_}NH\text{-}NH\text{-}C_6H_5$$

$$H+\ H+$$

and that then the repulsion of the two positive charges causes the molecule to break up between the two hydrogen atoms, whereupon the two portions again unite in such a manner that the positive charges are at a greater distance from each other. This representation accounts for the presence of [HCl]² in the equation of velocity, as according to this equation one mol. of hydrazobenzene reacts with two H-ions.

It is implicit in the quote above that the rearrangement is intramolecular, which had been recognized from the products obtained from various asymmetric

[132] A.F. Holleman and J. Potter van Loon, The transformation of benzidine. *KNAW Proceedings.* 6:262-267 (1904).

hydrazobenzenes. Dewar was the first to try to explain that in 1946.[133] Substantially more work was done in the 1960-1980 period as orbital symmetry became understood.

16. An Electronic Theory of Chemical Bonds

From the publications of van't Hoff and Le Bel (1874) until the publication and comprehension of Erwin Schrödinger's wave mechanics (1926-1930), organic chemist built on a structural theory of chemistry without any tie to the first principles of physics. Fisher's work with stereochemistry of sugars and the success of some of the early reaction mechanism work (e.g., Sn2 inversion of absolute configuration) provided evidence that the structures actually existed, but there was no physical law that rationalized the structures or the stability (inertness) of the structures. Lewis had proposed that atoms preferred to have 8 electrons located at the corners of a cube (1902) and that

---

[133] M. J. S. Dewar. The kinetics of some benzidine rearrangements, and a note on the mechanism of aromatic Substitution. *J. Chem. Soc.*, 1946, 777-781.

configuration could be rationalized by electrostatic repulsion of the electrons. In 1916, he introduced the idea of electron pair bonds. However, the valence shell electron pair repulsion (VSEPR) model was not introduced until 1957. Irving Langmuir (1881–1957) was the first to describe the covalent bond (1919).[134] This was a concept that was clearly needed to rationalize structural chemistry. But organic chemists were moving ahead of theoretical chemists with the intuitive applications of structural chemistry.

Organic Chemists Go Where Others Fear to Tread

Johannes Thiele (1865–1918) was probably the first chemist to try to rationalize reaction of conjugated pi-bond systems. In 1899,[135] he was faced with the fact that in 1,3-budadiene systems, reaction (e.g., bromination) often occurred at the 1 and 4 positions.

---

[134] I. Langmuir. The Arrangement of Electrons in Atoms and Molecules. *J. Amer. Chem. Soc.* 41 (6): 868–934 (1919).

[135] J. Thiele. *Zur Kenntnis der ungesättigten Verbindungen* (On our knowledge of unsaturated compounds). *J. Liebig's Annalen der Chemie.* 306: 87-266 (1899).

He interpreted this as formation of saturated valences on the two center carbons (i.e., 2 and 3) with partial valence[136] associated with positions 1 and 4. He extended the partial valence concept to a nearly modern interpretation of benzene (avoiding the implication of oscillation between two 1,3,5-cyclopentatrienes). He also applied these analyses to explain substitution patterns on benzene and other systems such as fulvenes (prepared by the reaction of aldehydes and ketones with cyclopentadiene).[137]

$$CH_2$$

HC   CH

HC—CH

Drawing by Yikrazuul, source Wikimedia Commons

In Thiele's system, the electronic structure was essentially static (much like the modern molecular orbital image of molecules).

---

[136] By the 1950s, the term 'free valence index" had entered the vocabulary calculated by the Hückel molecular orbital method.

[137] J. Thiele, J. *Uber Ketonreactionen bei dem Cyclopentadiën. Berichte der deutschen chemischen Gesellschaft* 33 (1): 666–673 (1900).

In 1909, Lapworth found himself at the University of Manchester with Robert Robinson (1886–1975), a student of W. H. Perkin, Jr. They became friends and communicated constantly until 1912 when Robinson finished his degree and obtained a position at the University of Sydney. Robinson returned to Britain in 1915 (University of Liverpool). In 1916[138], Robinson wrote his first paper that incorporated ideas of electrons that was similar to Thiele but showed obvious influence from Lapworth.

After several changes in his career, Robinson arrived at St. Andrews in 1921. Drawing from the new ideas of bonding by Lewis and Langmuir (1919), Robinson was one of the first chemists to attempt to rationalize chemical reaction mechanisms in terms of *movement of electrons* (i.e., making and breaking bonds). This system is the basis of valence bond concepts. In 1922, Robinson rejoined Lapworth at Manchester. They obviously communicated, but published separately; and in 1922, both introduced "curly arrows" to indicate electron movement. However, Robinson's use of this notation led to our modern practices, while Lapworth's application was more abstract. Robinson drew bonds

---

[138] E.E.P. Hamilton and R. Robinson. *J. Chem. Soc.* 1916:1029

that were in the process of breaking or forming as dashed lines *in the transition state*[139]:

> *"The necessary condition precedent to chemical change is the activation of one or more molecules taking part in the reaction; this is done by cohesion and rearrangement of valencies, most probably synonymous with changes in the position of electrons.*
>
> *. . . The activated molecules are further assumed to be polarized and to contain partially dissociated valences.*
>
> *. . . The representation of the phenomenon of conjugation and addition to conjugated systems is much simplified by the use of the theory of divisible and polar valency."*

His induced polarizations (following ideas of Lapworth) suggest the direction of electron movement.[140] His focus was on formation of a complete octet, which we would now likely interpret as an effect of electronegativity, i.e., electronegative

---

[139] Robinson. R. Manchester Memoirs. 64.4 (1920). This quote is taken from M.D. Saltzman. The Robinson-Ingold Controversy. *J. Chem. Ed.* 57(7):484-488 (1980).

[140] W.O. Kermaek and R. Robinson. An explanation of the property of induced polarity of atoms and an Interpretation of the theory of partial valencies on an electronic basis. *J. Chem. Soc.* 1922.427.

elements polarize bonds as they attempt to satisfy their octet (2s and 2p or 3s and 3p).

Concurrent with the work of Lapworth and Robinson, there was another scientist working outside the normal academic structure. Bernard Flurscheim (1874-1955) has been described as an "enigmatic figure", but his career was not much different that Charles Darwin or Arthur Michael. He was born to money in Germany and took an interest in chemistry. He studied with several chemists in Europe including Thiele and married a British woman in 1905. Thus, he settled in Britain; and rather than vie for academic chairs, he simply built himself a laboratory where he worked on projects that interested him. In the early 1900s, he extended the ideas of Thiele and was apparently influenced in turn by Lapworth and Langmuir. In particular, he translated the idea of completing the octet into the idea of maximizing valences: an atom with unsaturated valence drew electrons from neighbors and an atom with saturated valences released electrons to neighbors. This concept was successful in predicting electrophilic substitution patterns on aromatic systems and Flurscheim was a regular participant in discussion at the Chemical

Society of London, where he met and befriended Christopher Ingold (1893–1970).

Ingold was confronted with the Lapworth-Robinson interpretation of alternating charges[141] and the Flurscheim idea of alternating valences. At the time, the difference in single (sigma only) and double bonds (sigma and pi) was not understood. Today, I think we would say that Robinson was looking at sigma-effects and Flurscheim was looking at pi-effects. In any event, Ingold hit on the idea that the nitroso substituent (-N=O) would allow him to test the theories in the reactivity of nitrosobenzene. Robinson would predict nitroso to be a meta-director (based on alternating polarity starting with oxygen as negative) while Flurscheim would predict that it would be an ortho/para-director based on conjugation. In 1925, Ingold tested the hypothesis[142] and observed para-substitutions and no meta-substitution with several electrophiles (Cl^+, Br^+, $^+NO_2$). From this, he concluded that the Lapwoth-Robinson model was incorrect. The results were communicated at a meeting of the society

---

[141] Theories of organic chemistry by Ferdinand August Karl Henrich. 1922. See pp. 33-107.

[142] C.K. Ingold. *J. Chem Soc.* 1925:513.

before they appeared in print and Robinson had his rebuttal in print[143] before Ingold's paper appeared. Remember that the distinction between a sigma and pi bond did not exist at that time[144], so Robinson's rebuttal took on a semi-empirical nature comparing conjugated systems to the crotonoid system. Ingold countered with an example of (alpha-methoxyvinyl)benzene, which yielded ortho/para-products in spite of the fact that the alternating charge theory would predict meta-substitution.

By this time (mid-1925), there was acrimonious allegations that Ingold was misinterpreting what Robinson's theory predicted. In truth, Robinson does not seem to have had a comprehensive theory. He was just supporting the earlier ideas of Lapworth and adding *ad hoc* arguments about conjugation. Ingold's predictions based on Flurscheim's ideas (which were

---

[143] R. Robinson. *J. Chem. Ind.* 43:2397 (1924) and 44:456 (1925).

[144] Today the explanation is straight forward: Although the nitroso group withdraws electrons through the sigma system and deactivates the ring towards electrophilic reactions, the lone-pair on nitrogen feeds electron density back into the pi system and directs substitution especially to the para-position by conjugation. The ortho-position is deactivated to the point that it is not a primary site for electrophilic attack.

also not all that clear) appeared in some respects to be *post hoc* rationalizations of experimental observations. Ingold attempted to drive the last nail into the alternating charge theory (of Lapworth) with studies involving benzylamines and alcohols. Ingold's interpretation of his results were reasonable (*J. Chem. Soc.* 1925:1800)[145]:

> *"...We suggest, therefore, without denying the possibility of polar alternation, that the facts thus far adduced constitute grounds for the conclusion that the propagation of alternating unsaturation is the prime directive process in ordinary aromatic substitution."*

But, his reference to "the possibility of polar alteration" appeared condescending to Robinson, who took offense. Robinson's defense was interesting in that he seems to have dropped the alternating polarity concept in favor of an inductive effect on top of a conjugation effect. Robinson (really for the first time) articulated a comprehensive theory in 1926.[146] He provided a pre-print to Ingold, who responded graciously with

---

145 Quoted from M.D. Saltzman. The Robinson-lngold Controversy. *J. Chem. Ed.* 57(7):484-488 (1980).

146 J. Allen, A.E. Oxford, R. Robinson and J.C. Smith. *J. Chem. Soc.* 1926:401.

complements. Moreover, in his next paper (1926), Ingold applied Robinson's concepts and curvy arrows symbolism, but cast them as a clear articulation of the Flurscheim concepts of free and bound affinity.

An interesting problem then arose. The highly contentious work by Ingold involving benzylamine was proven by Robinson to be wrong.[147] It was not evidence against the Lapworth-Robinson concepts. One might wonder how Ingold made this mistake. Today, such a mistake would be unthinkable, but in 1925, the available physical techniques apparently excused Ingold's error... nonetheless, his desire to undo Robinson might have been a factor in the way he interpreted his data.

Until 1928, Robinson and Lapworth were frequent companions although Robinson's work was primarily involved with synthesis of natural compounds. Robinson published little else on the subject; but after he moved to University College (1930), Ingold flooded the popular journals (e.g., a *Chemical Review* article in 1934) with examples and applications of these concepts (conjugation and induction) first articulated by Robinson. Ingold's nomenclature (e.g., electrophile

---

[147] H.R. Ing and R. Robinson. *J. Chem. Soc.* 1926:941.

and nucleophile) systematically displaced Robinson's terminology. This was seen by Robinson, with some justification, as Ingold laying claim to Robinson's ideas. *This story will be continued.*

In the next chapter, we will see the redemption of the organic chemists by advances in atomic physics.

IV. Early-Modern Organic Chemistry (1930-1955)

The early-modern period of organic chemistry was initiated by the clarification of the electron configuration of atoms and Schrodinger's models of atomic orbitals. From here, organic chemists could use three-dimensional structural chemistry to explain and predict the mechanism of chemical reactions complete with stereochemistry of the reactants and products. The petroleum industry became the source of an ever-growing supply of hydrocarbon fuels and feed stocks. Polymer chemistry and the manufacture of synthetic fibers and plastics was a hallmark of this period. This period was punctuated by World War II (1939-1945) during which there was intensive technological development in Germany (synthetic fuels) and the US/Britain (alkylation of isooctane, synthetic rubber, penicillin, chloroquine, DDT, ion chromatography, x-ray crystallography) including instrumentation. But, during the war this research was generally kept secret.

Thus, in 1946-1948, there was an outpouring of technical data and knowhow into the academic community. It took several years for war-time knowledge to be digested, consolidated and produce major new breakthroughs. Concurrently, multi-step organic syntheses of natural products with stereo-chemical control was introduced. Although the syntheses had no commercial value, they confirmed natural structures and helped reveal subtle elements of reaction mechanisms controlled by quantum mechanics. I view the determination of the structure of DNA (by Watson and Crick using the modeling techniques provided by structural chemistry and the x-ray crystallography data of Franklin) as the culmination of this period.

1. Electronic Theory and the Basis of Structural Chemistry

By 1930, organic chemists had fully accepted the hypothesis that chemical compounds were composed of atoms bound to one another in specific inert arrangements (radicals) and this hypothesis was enormously productive. But, if Hermann Kolbe had still been alive, he could still have complained that there was no physical law that accounted for the observations. This odd situation was to change between 1930 and 1955.

The Physicists:
Rutherford, Bohr, Schrodinger and Heitler[148]

The practical contributions of organic chemistry had caused it to run far ahead of physical theories of atoms and molecules from 1870 to 1930. Indeed, some chemists (like Kolbe) had doggedly relied on the physical theory of pure electrostatics; and thus, had missed the wave of structural organic chemistry (1875-1900).

Nonetheless, without reliance on any chemical observations, Rutherford (based on the work of Geiger) realized that most of the mass and positive charge of atoms was concentrated in a tiny (10^{-15} m radius) nucleus and most of the volume of the atom (10^{-10} m radius) was filled by the negative electric charge of a swarm of electrons. This structure of the atom could not be rationalized by classical physics (e.g., electrostatics). Bohr then explained the atomic

---

[148] I discussed the contributions of Rutherford, Bohr and Schrodinger in some detail in my book *Matter and Energy and informal history* (2014) available on Amazon/kindle.

structure and atomic spectra in terms of distinct energy levels for electrons that occupied stable orbits around the nucleus. De Broglie invoked the wave-particle duality of electrons and turned Bohr's circular orbits into orbitals (volumes of space where electrons were most likely to be found).

Heisenberg managed a very abstract mathematical representation of the orbitals that was even difficult for the physicists to grasp. But Heisenberg's method was immediately displaced by Schrodinger's introduction (1926) of a physical representation of an atom in which the orbitals were described by wave functions (Ψ(n, l, ml)) of negative charge, which are derived from spherical harmonics. Spherical harmonics have very important characteristics: non-spherical shapes determined by (quantum number l) and directionality determined by (quantum number m l).

In my opinion, Pauli (1925) was given far too much credit for the idea that *no more than two electrons* (with opposite magnetic moments, s = +/- ½) *can occupy* the same orbital. The reverse of Pauli's concept,[149] i.e., *that two negatively charged electrons <u>can</u> occupy the same volume of space in spite of their electrostatic repulsion* leads

---

[149] not exploited by Pauli

directly to the idea that *electron pairs* (which are critical to the understanding of chemical bonding and stereochemistry) are able to occupy a volume of space by balancing electrostatic repulsion with electromagnetic attraction. At best, Pauli confirmed Lewis's ideas of electron pairs.

Fritz London (1900–1954) and Walter Heitler (1904–1981) were young Jewish physicists in a Germany that was rapidly becoming anti-sematic in the late 1920s. London had done some original work on forces between atoms attributable to induced polarization of their electric fields (i.e., dispersion forces). Heitler in particular was thoroughly immersed in the wave of atomic physics developed in Europe during the 1920s.

While working in Zurich, Switzerland with Schrodinger, they followed up Schrodinger's analysis of atoms with a similar analysis of molecules (beginning with H_2) and out of this work, the idea of the *covalent bond of shared electron pairs* (homo-polar bonding) was born.[150] This led to the generalized

---

[150] W. Heitler, F. London. *Wechselwirkung neutraler Atome und homöopolare Bindung nach der Quantenmechanik. Zeitschrift für Physik* 44 (6–7): 455 (1927).

valence bond theory. And, this is where the physicists left the problem in 1930.

Linus Pauling and Hybridization

The chemists were not satisfied with the atomic theory as it stood in 1930, because it still did not explain what they knew to be true in Langmuir's 1919 paper.

Linus Carl Pauling (1901–1994) overcame a humble childhood with enormous effort and ingenuity to receive a PhD in physical chemistry in 1925 from the California Institute of Technology. The interpretation of atoms and electrons by Schrodinger and Pauli were on the table, but their description did not satisfy the needs of chemists. Pauling received a Guggenheim Fellowship (1926-1927) to study in Europe under Sommerfeld, Bohr and Schrodinger. During his stay in Zurich, Pauling also met Heitler and London who were in the process of developing the basic valance bond theory (i.e., covalent bonds are formed by overlap of atomic orbitals into which each atom contributes unpaired valence electrons).

When he returned to the US, Pauling was probably the only chemist in the world who fully grasp quantum mechanics. But, his next five years were focused primarily on the stability of ionic crystals based on the charges and relative radiuses of the ions.

After another trip to Europe in 1930, he focused more on electronegativity (1932) and the conceptual transition of an ionic bond (with large differences in electronegativity) to a polar bond (unequally shared electrons between atoms with similar electronegativities) and non-polar covalent bonds (e.g., C-C). With this basis, he took Schrodinger's spherical harmonics (atomic orbitals) and Heitler's valence bonds and made something very useful for chemists. Pauling realized that the s-type orbital and the *set of* three p-type orbitals were spherically symmetrical, but if he blended[151] them together to obtain four sp³ hybrid orbitals, he could rationalize tetrahedral symmetry around carbon (and other elements Li, Be, B, C, N, O, F, Ne). Suddenly everything fell into place. The covalent bonds between atoms involved overlap of orbitals from different atoms. Sometimes these were hybridized to place the electron density directly between the nuclei (sigma bonds). The sigma bond angles of 109.5°, 120° and 180° could be explained by hybridization of the s

---

[151] Blending the orbitals (2s and 2p) together this way was rationalized because they were similar size and energy.

orbital with three p-orbitals (sp^3), two p-orbitals (sp^2) or one p-orbital (sp), respectively. Pauling refined his ideas in the 1930s and published his widely cited book *The Nature of the Chemical Bond* (first edition, 1939).

Pauling was not the only scientist working in this field. In 1927-28, Robert Mulliken (1896–1986) and Friedrich Hund (1896–1997) were extending the same ideas into the molecular orbital (MO) theory in which (hybridized) atomic orbitals were combined to make molecular orbitals. This idea came into full fruition in 1929 with John Edward Lennard-Jones (1894–1954) and his method of linear combinations of atomic orbitals (LCAO).[152]

It gradually became apparent that the hybridization around atoms, which had (non-bonding) lone-pairs of electrons was not perfectly as expected. In 1957, this led to the introduction of the concept of valence shell electron-pair repulsion theory (VSEPR).

---

[152] J. E. Lennard-Jones. The electronic structure of some diatomic molecules. Trans. Faraday Soc. 25:668-686 (1929).

Sigma and Pi Bonds

After earning his doctorate, Erich Hückel (1896–1980) worked with Peter Debye (1884–1966) in the 1920s to find a way to account for electrostatic effects in electrolyte solutions. He then spent part of 1928-29 with Niels Bohr and undoubtedly followed the developing ideas of Pauling about hybridization. Hückel realized that after linear combination of atomic s and p orbitals to make hybrid sp² or sp type orbitals, there were "left-over" (un-hybridized) p-orbitals and the close proximity of the atoms allowed p-orbitals on adjacent atoms to overlap concurrently with the overlap of the hybridized orbitals. He thus deduced that the direct overlap of the hybridized (2s and 2p) orbitals of carbon (as well as boron, oxygen, nitrogen, etc.) created a (sigma) bond (symmetrical around the interatomic axis), which allowed free rotation; but the overlap of the 2p orbitals would is not symmetrical around the bond axis; indeed, it required a specific alignment, *which cannot be altered without breaking the (pi) bond.* The sigma/pi concept was expressed in 1930; and by 1931, Hückel had developed valance bond (VB) and molecular orbital (MO) descriptions of benzene. Unfortunately, he was not very good at communicating his ideas and most of the practical applications were developed decades later by others.

Structural organic chemistry finally had a rationale tied directly into physical phenomena.

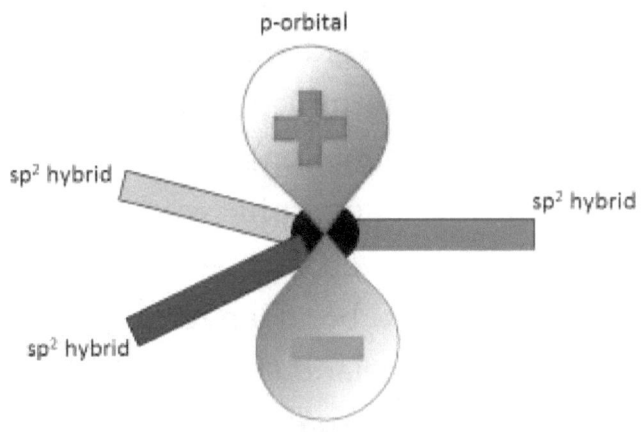

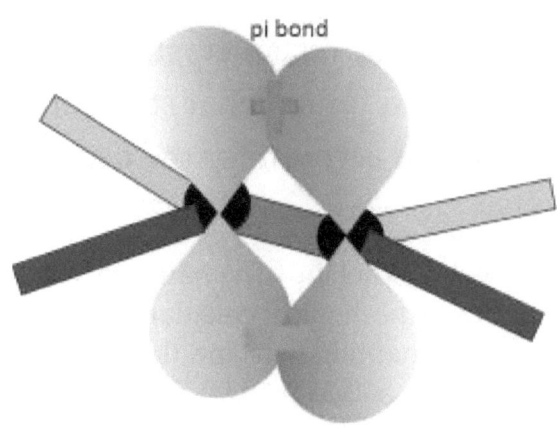

Aromatic Compounds and MO Theory

In 1930, Erich Hückel had extended the work of Linus Pauling to explain sigma and pi bonding and lay the groundwork of molecular orbital theory. However, Hückel did not carried the idea through to examine the unusual stability (as determined by heats of combustion and hydrogenation) of aromatic compounds. The crystallographic work of Kathleen Lonsdale 1928-1931 showed that benzene was a planar molecule with equal C-C bond lengths. In 1933, Pauling and George Wheland (1907-1962)[153] were the first to appreciate the unusual stability of benzene and calculate its "resonance energy" (i.e., stability relative to a hypothetical system of isolated double bonds).[154] Without reference to the recent crystallographic data of Lonsdale, they argued for a planar structure with resonance that would account for the fact that no one

---

[153] L. Pauling and G.W. Wheland. The nature of the chemical bond. V. The quantum-mechanical calculation of the resonance energy of benzene and naphthalene and the hydrocarbon free radicals. *J. Chem. Phys.* 1: 362-374 (1933).

[154] Interestingly, they do not cite Lonsdale's crystallographic results.

had ever isolated but one isomer of ortho-substituted benzenes. I have reproduced their Table 1 as follows:

Table 1			
	Total Energy	Resonance Energy	a/b
Single Kekulé structure	$Q + 1.5\alpha$	0	
Resonance between Two Kekulé structures	$Q + 2.4\alpha$	0.9α	$1 : 0$
Resonance among all five Kekulé structures	$Q + 2.6055\alpha$	1.1055α	$1 : 0.4341$
a/b is the ratio of the coefficient of structures A and B to that of the singly excited structures C, D and E.[155]			

Using thermodynamic data[156] Pauling calculate the value of α:

---

[155] Structures A and B are the Kekulé structures and C, D, E were drawn as flat "Dewar benzene" structures, but should be interpreted as three para-diradicals.

[156] L. Pauling and J. Sherman. The nature of the chemical bond. VI. The calculation from thermochemical data of the energy of resonance of molecules among several electronic structures. *J. Chem. Phys.* 1: 606-617 (1933).

α = -1.5 eV (1 eV = 96 kJ/mole = 23 kcal/mole)

Alpha (α) is called the "Coulomb integral" and is interpreted as the energy lost by an electron moved from infinite separation to an un-hybridized p-type orbital on a carbon atom.

In his next paper (*The nature of the chemical bond. VI*), Pauling and Sherman raise the resonance stabilization of benzene to 1.62α (156 kJ/mole)[157] and provide the following table:

Table II				
Compound	Formula	E observed	E′ calculated	Resonance energy
Benzene	C_6H_6	58.20	56.58	1.62α
Toluene	$C_6H_5CH_3$	70.58	68.88	1.70α
Ethyl-benzene	$C_6H_5C_2H_5$	82.90	81.18	1.72α
Propyl-benzene	$C_6H_5C_3H_7$	95.27	93.48	1.79α

These data are convincing that the stabilization is associated with the electronic structure of the benzene ring rather than the side chain.

---

[157] The modern value is 152 kJ/mole.

Hartree-Fock method

Douglas Hartree (1897-1958) was a student of Ernest Rutherford and received his PhD in 1926. As soon as Schrodinger published his wave mechanics paper in 1927, Hartree made the assumption that the field around a free atom must be spherically symmetrical; and using the principle that the orbital computed for the atom must conform to this spherical symmetry, he developed a calculation procedure known as the self-consistent field (SCF) model. Recall that in the hydrogen atom (i.e., a single electron) the energy of the wave functions were determined only by the principle quantum number (n), but in real (i.e., multiple-electron) atoms there must be repulsions among the electrons and Gauss's law indicates that the field experienced by an electron at a some distance from the nucleus is determined by the net enclosed charge (screening constant). These concepts were initially included as an empirical "quantum defect," but Hartree wanted to do energy and electron distribution calculations from first principles (*ab initio*).

In 1928, John Slater (1900-1976) provided an important tool by applying a perturbation method to the hydrogen-like atomic orbitals to obtain approximate values of the integrals.[158] In 1930, Vladimir Fock (1898–1974) showed that incorporation of Slater approximations solved some of the problems with the Hartree model. Hartree worked out a method of applying the system by 1935. By 1938, Charles Coulson (1910–1974) had done one of the first *ab initio* calculations <u>by hand</u> for H_3^+, but the computations using numerical analysis (since exact solutions were impossible) were too burdensome to carry out for anything more than a small atom until automatic computing became available in the early 1950s.[159]

---

[158] J.C. Slater. The self-consistent field and the structure of atoms, *Physical Review*. 32:339-348, (1928).

[159] Born and Oppenheimer had proposed a simplifying assumption in 1927, i.e. assume that the electronic and the nuclear movements can be calculated separately (because the electrons move much faster), but this was not enough to allow computations before 1950. See also the Franck–Condon principle.

2. Reaction Mechanisms and Physical Organic Chemistry

Lapworth and Ingold had begun the study of organic reaction mechanisms between 1900 and 1930. As the electronic theory of chemical bonding matured in the 1930s, the discipline of "physical organic chemistry" began to take shape beyond the historical focus on synthesis.

Empirical Relationships Based on Electronic Structure

Our understanding of chemical equilibria goes back to 1864[160] and by 1916 the Henderson-Hasselbalch equation had been developed to relate the concentration of protons in solution [H+] to a thermodynamic parameter pKa:

$$pH = pKa + \log\left([A^-]/[HA]\right)$$

---

[160] See review by J. Reijenga, A. van Hoof, A. van Loon and B. Teunissen. Development of Methods for the Determination of pKa Values. *Analytical Chemistry Insights*. 2013:8 53–71.

Note that the pKa = pH when the acid (AH) is 50% ionized. But this equation is only valid in dilute solutions. For concentrated solutions, the concentration terms need to be corrected by activity coefficients.

In 1923 Johannes Brönsted (1879-1947)[161] and Thomas Lowry (1874–1936)[162] independently published the general theory that acids are proton donors and bases are proton acceptors. Brönsted followed up his proposal with several papers (1924, -26, -27, -28) culminating in his article in *Chemical Reviews*.[163] These papers developed ideas about the thermodynamics of acids, activity coefficients, and the factors that determine their strength.

Concurrently, James Conant (1893–1978) finished his doctorate at Harvard (1916) and started a chemical manufacturing company to take advantage of shortages (e.g., benzoic acid) caused by the WWI embargo of Germany. His detractors point out that his decisions in this period (including whom to marry) were driven by

---

[161] May 4, 1923 in *Receueil des Travaux Chimiques des Pays-Bas*.

[162] January 19, 1923 in *Chemistry and Industry*.

[163] J.N. Brönsted. *Chem. Rev.* 5:231 (1928).

self-interest and a desire to return to Harvard. Regardless, he was recalled to Harvard shortly after the plant was established in New Jersey; and as his partners including a well-known Harvard football player (Stanley Pennock), began to start up production (November 27, 1916) there was an explosion that killed the football player and two technicians. Naturally, some people felt that Conant might have some responsibility for the accident.

Conant was employed by Harvard, but was inducted into the US Army Sanitary Corps as a major (September 22, 1917) and moved to American University (in NW Washington, DC)[164] where he worked on chemical weapons (poisonous gases) and helped develop Lewisite. This reaction involves an electrophilic addition of $AsCl_3$ to acetylene:

$$AsCl_3 + C_2H_2 \rightarrow ClCH=CHAsCl_2$$

---

[164] Incidentally, the waste from these experiments were buried in pits further out Massachusetts Avenue in an area that was later developed as "Spring Valley" with numerous expensive homes. The homeowners started discovering these discarded chemical wastes *circa* 1990 and the US Army Corps of Engineers spend several years tracking down and removing the waste.

Lewisite

Drawing by JaGa, source Wikimedia Commons

After the war, Conant became an associate professor at Harvard[165]; and in 1924-25, he published three quantitative studies on the rates of reaction of alkyl chlorides and bromides with sodium iodide in acetone (i.e., the Finkelstein halogen exchange reaction). This is a Sn2 reaction facilitated by the solubility of sodium iodide in acetone in which sodium bromide and sodium chloride are insoluble. In 1925, he toured chemistry laboratories in Europe (1925). When he returned, he settled at Harvard and engaged in very prolific research between 1927 and 1932. This research followed the recently published acid theory of Brönsted. But, instead of using water, he used glacial (100%) acetic acid as solvent.[166] This methodology

---

[165] Louis Fieser (1899–1977) was a graduate student of Conant at Harvard and received his PhD in 1924. During WWII, Fieser invented napalm (jellied gasoline) used as fire bombs.

[166] N.F. Hall and J.B. Conant. A Study of superacid solutions. I. the use of the chloranil electrode in glacial acetic acid and the

began the idea of "super acids" (e.g., non-aqueous media). This work included work that later supported the alkylation of olefins.[167]

Of course, pKa depends upon temperature and ionic strength. Nonetheless, at constant temperature and ionic strength, pKa can be related to the enthalpy of the reaction *if (and only if)* the enthalpy term (ΔH) is much larger than the entropy term ($T\Delta S$) following the van't Hoff approach:

$$d(\ln Ka)/dT = \Delta H/RT^2$$

In a very gracious acknowledgement, in 1966, Louis Hammett (1894–1987) explained:

> "The parent of all relationships of this kind is the discovery by Bronsted and Pedersen of general acid and base catalysis and of the rule that the logarithms of the rate constants of the catalyzed reactions are

strength of certain weak bases. *J. Amer. Chem. Soc.* 49(12):3062–70 (1927).

[167] J.B. Conant and G.W. Wheland. The Structure of the Acids Obtained by the Oxidation of Tri-isobutylene. *J. Amer. Chem. Soc.* 55(6):2499–2504 (1933).

linearly related to those of the acidity constants of the catalyzing acid or base. ..." [168]

Hammett was referring to the equation that he developed in the 1930's with H.L. Pfluger:[169]

$$\log k = \log k_0 + \sigma\rho$$

In this equation, the rate constant for a reaction (k) is related to some arbitrary standard (k_0) by the factors sigma and rho. Hammett (1937) explains that this equation is derived from a more fundamental equation:

Relative $\Delta G^{\ddagger}$ = RT (ln k – ln k_0)

$$= RT \ln(k/k_0) = (A/d^2)\{(B_1/D)- B_2\}$$

Where

d = distance from the substituent from the center of reaction

---

[168] Quote from J. Shorter. The prehistory of the Hammett Equation. *Chem. Histy.* 94:210-214 (2000).

[169] L.P. Hammett and H.L. Pfluger. *J. Am. Chem. Soc.* 55:4079 (1933). L. P. Hammett. Some relations between reaction rates and equilibrium constants. *Chem. Rev.* 17(1):125–136 (1935).

L.P. Hammett. The effect of structure upon the reactions of organic compounds. Benzene derivatives. *J. Am. Chem. Soc.* 59(1):96-103 (1937).

D = dielectric constant of the medium

A, B_1 and B_2 are constants that are independent of temperature and solvents (as long as the reaction mechanism does not change)

The first term is dependent only on the substituent and its position relative to the center of reaction (i.e., the carbon where bonds are being made and broken) and B_1 and B_2 depend only of the nature of the reaction. Thus,

$$\sigma = A/2.303R \text{ and } \rho = (1/Td^2)\{(B_1/D)- B_2\}$$

Using this empirical equation, numerous data sets have been developed for the sigma and rho parameter. These Hammett relationships have predictive value and help us understand the mechanisms of reaction. Because equilibrium and rate constants can be related to free energy and free energy of activation, respectively:

$$\Delta G = -RT \ln K_{eq}$$

And the Arrhenius equation (1899):

$$k = A \exp[-\Delta G^{\ddagger}/RT]$$

they are called *linear free energy* relationships.

Aliphatic Substitution and Elimination Reactions

Ingold moved to University College, London in 1930. In that same year, Ingold was an "outside examiner" for the thesis defense of Edward Hughes (1906-1963) at University College NW, Bangor (Wales). Ingold apparently liked what he heard because he brought Hughes with him to London where they worked together until 1943 (when Hughes was appoint chair at Bangor).

Ingold and Hughes went about trying to systematize organic reactions (for example, by classifying reactions of various nucleophiles under the same category such as nucleophilic substitution at an aliphatic carbon).[170] They did this to bring all the mechanisms together, rather than organize by types of compounds (e.g., alcohols, halides, nitriles, sulfides).

The well-known (but mysterious) phenomenon of the "Walden inversion" had been first documented in

---

[170] J. H. Ridd. Organic Pioneer. *Chemistry World*. December 2008:50-53.

1896[171], but lacked an explanation in 1930. Fischer had done extensive work and hypothesized that an addition preceded the elimination of the leaving group in these reactions[172], but the issue was not resolved. For example, Muskat attempted to explain it in 1934[173], but he assumed that retention of configuration involved retention of all the electron pairs with the asymmetric carbon; while inversion of configuration involved loss of an anion. Both mechanisms were drawn essentially as first-order reactions. Between 1932 and 1935, Hughes and Ingold began studying the kinetics of

---

[171] P. Walden. *Uber die gegenseitige Umwandlung optischer Antipoden* (Concerning the inter-conversion of optical antipodes [isomers]). *Berichte der deutschen chemischen Gesellschaft.* 29(1):133–138 (1896).

[172] See for example the discussions on pp. 185-195 in *Theories of Organic Chemistry* by Ferdinand August Karl Henrich (1922).

Fischer introduced the use of ball-and-stick models to demonstrate his idea, and showed a five-coordinate carbon, but he did not conclude that the approach of the nucleophile was directly opposite to the leaving group.

[173] I.E. Muskat. The Mechanism of Walden Inversion in Sugars: The Inversion of p-Toluenesulfonyl Esters of Rhamnose. *J. Am. Chem. Soc.* 56(12):2653–2656 (1934).

nucleophilic substitution reactions and deduced that there were two distinct mechanisms: first order (Sn1) and second order (Sn2). They then were able to associate the Walden inversion to the Sn2 mechanism and explain why Sn1 resulted in loss of optical activity.[174]

Drawings by Calvero, source Wikimedia Commons

As (I assume) everyone reading this document is aware, these mechanisms readily account for randomization of configuration (Sn1) and inversion of configuration (Sn2). Of course, Ingold and Hughes

---

[174] W.A. Cowdrey, E. D. Hughes and C. K. Ingold. Reaction kinetics and the Walden inversion. *Nature* 138:759-759 (1936).

noticed that elimination frequently accompanied substitution and simultaneously developed the E1 and E2 mechanisms through the 1930s.

3. Synthetic Rubber

Into the 1930s, natural rubber continued to be the essential material of industry. The process invented by Lebedev for making butadiene (from ethanol) and polymerizing it with sodium (Synthetic Kauchuk, CK-1) flourished only in Russia. In Germany, Eduard Tschunkur (1874-1946) invented Buna[175]-S (styrene-butadiene rubber); and in 1930, he and Erich Konrad (1894-1975) developed Buna-N (copolymer of acrylonitrile and butadiene) rubber, which were patented by I.G. Farben (1934). I.G. Farben also manufactured polyisobutylene (PIB) in 1931. These products all used the Lebedev sodium technology.

In the US, Wallace Carothers (1896–1937) was hired by DuPont in 1928 and worked with vinyl acetylene (obtained by dimerization of acetylene). When exposed

---

[175] "Buna" stands for butadiene-natrium (Na, sodium).

to hydrochloric acid, vinyl acetylene produces 4-chloro-1,2-butadiene, which rearranges in the presence of copper(I) chloride (CuCl) to 2-chlorobuta-1,3-diene (chloroprene).[176] Chloroprene was first polymerized in 1930 and was marketed in 1931 as DuPrene™. The product was significantly improved after the process was modified and then marketed under the name Neoprene™. Neoprene was more resistant that natural rubber to organic solvents and oils and found its way into applications such as flexible fuel lines and pump diaphragms for fuels and organic solvents.

In the U.S. (1937), William Sparks[177] (1905-1976) and R. Thomas managed to make an elastomer from a diene and co-monomer (e.g., isobutylene with isoprene) using cationic polymerization; and this became the basis for butyl rubber in the US used for inner tubes (but it was not durable enough for tires).[178]

---

[176] This process has been replaced by chlorination, followed by rearrangement to 3,4-dichloro-1-butene and elimination of HCl with base.

[177] President of the American Chemical Society in 1966.

[178] First Standard Oil plant in Baton Rouge in 1941.

Germany and the USSR were always on the lookout for technology for synthetic materials because of their limited natural resources and limited merchant marine. Early in WWII, the Germans received natural rubber from India (via the USSR) and Indonesia (via Japan and the USSR) to supplement their synthetic rubber production. In contrast, the US (which consumed about half of the world's supply of natural rubber) did not see synthetic rubber as an urgent priority until Japan invaded Indonesia in 1941. Under wartime stress, the U.S. government formed partnerships among companies[179] and academia to solve the rubber problem, which had to be resolved within 18 months[180] or the war might be lost. The technology selected was based on Buna-S for which the US was well prepared to provide the monomers from domestic oil. The government standard (GR-S) rubber was butadiene

---

[179] Standard Oil supplying butadiene and, The Firestone Tire & Rubber Company (Bridgestone/Firestone, Inc.), The B. F. Goodrich Company, The Goodyear Tire & Rubber Company, and United States Rubber Company (Uniroyal Chemical Company, Inc.).

[180] In 1939-41, the US had stockpiled about one million tons of natural rubber and a consumption rate of 500,000 tons per year were projected.

(75%) and styrene (25%), polymerized in emulsion (soap, water) with potassium persulfate (initiator) and dodecyl mercaptan (modifier). Pilot scale production started by April 1942 and the first production plants were on line in 9 months (autumn of 1942). These plants continued in operation through the Korean War (1953).

4. High-Compression Engines and Gasoline

The supercharger[181] was invented and adapted originally to allow commercial airplanes to fly to higher altitudes (e.g., coast-to-coast in the US over the Rocky Mountains) where the atmospheric pressure was lower, but it was soon realized that super-charging at sea level produced greater power and efficiency. This has

---

[181] Essentially an air compressor driven off an internal combustion engine to compress the charge of air going into the cylinder. General Electric engineers, led by Dr. Stanford N. Moss, set out to develop a supercharger. In a famous series of tests in 1917 and 1918, they dragged a 350-hp Liberty airplane engine to the top of Pike's Peak (14,109 feet) to conduct test with a turbosupercharger. Turbosupercharger refers to a compressor driven by the exhaust gases of the engine rather than by mechanical connections (such as gears).

obvious value in military airplanes. However, when the supercharged engine designs went over an effective compression ratio of about 8 to 1 (the mechanical displacement compression ratios were about 6.5 to 1), adding 3 mL TEL *per* gallon of gasoline (the highest concentration considered safe because of the toxicity of TEL) would not raise the octane of straight run gasoline (octane about 55-60) adequately.

It was clear by 1930 that petroleum refiners needed to improve the octane quality of their basic hydrocarbon product. Stopgap measures such a isolating the "heart cut" of low boiling isopentane (octane 90) from the straight run fuel were tried, but this component was only about 15% of natural petroleum, and once it was removed, the remaining naphtha had a pitifully low octane of about 30, which made it almost useless even as a motor fuel.

In March 1935, the results of experiments with 100-octane fuel at Wright Field with the Boeing P-26 were published in the *Journal of Aeronautical Sciences* under the title "Aircraft Engine Performance with 100-Octane Fuel." In November 1936, the Army finally convened an expert panel under Col. J. T. McNarney to examine the issue of standardizing military octane requirements. The committee included representatives of engine makers Wright, Pratt & Whitney and United Aircraft; and fuel suppliers Shell Oil, Phillips Petroleum and Standard Oil. The committee resolved *"that 100-octane fuel would be as the standard for the Air Corps by January 1, 1938."*

Cracking, Reforming and the Houdry Process (1930-)

In the United States, petroleum was readily available and American research into higher octane gasoline was different from the work in Europe. The European developments after 1930 will be discussed after summarizing the American situation. The first major step, in making the petroleum refinery more than just a distillery, was taken by Eugene Houdry (1892-1962) who was born in Domont, France and moved to the U.S. in 1930. With the main interest of obtaining more liquid fuel from petroleum, he took the residue from petroleum boiling above 370°C at atmospheric pressure and vacuum distilled it. The liquid was not volatile enough to be used as fuel in ordinary engines, but he discovered that under these conditions, he could "crack" the large molecules into lower molecular weight molecules. This process also increased the olefin (C_nH_{2n}) content of the fuel.

$$C_xH_{2x+2} \rightarrow C_aH_{2a+2} + C_bH_{2b}$$

where x = a + b

It was also recognized that cracking increased gasoline's octane. During this same period (1933-1939), some "reforming" of straight chain aliphatics (C_nH_{2n+2}) into branched chain aliphatics was accomplished by rearrangement reactions that accompanied thermal cracking.

Houdry improved the un-catalyzed (*i.e.*, thermal) cracking process by finding catalysts that facilitate rearrangement and elimination reactions. Using the Houdry catalytic cracking process, fuel of about 90 octane could be obtained after addition of TEL. There was, however, a major problem with cracked fuels. The olefins in the cracked fuel tend to polymerize forming gum. The gum collected in the carburetor, which was a much more serious matter for airplanes than for automobiles. Thus, while cracked hydrocarbons became popular as component of "motor fuels," they were usually specifically prohibited from "aviation fuels." In some cases, the cracked fuels could be hydrogenated to eliminate the olefins, but this also lowered the octane number.

Cracking:

R-CH$_2$-CH$_2$-R' → R-H + H$_2$C=CHR' (olefin)

Hydrogenation:

H$_2$C=CHR' + H$_2$ →H$_3$C-CH$_2$R' (aliphatic)

Low molecular weight (C2 -C4) hydrocarbons were a byproduct of the Houdry cracking process. Initially, these gases were just another waste, but work was soon underway to find uses for them. The two most important components of this mixture were butane and butene.

Polymerization of Butenes to Manufacture Octane (1934-1941)

From the fractional distillations and particularly from the cracking processes (thermal and catalytic), oil refineries produced a large quantity of butane (C$_4$H$_{10}$) and butene (C$_4$H$_8$). These volatile gases were originally either "flared" (*i.e.*, burned at the end of an outlet pipe) as waste or burned in the refinery for fuel. About 1930 (even before Houdry's work), the Universal Oil

Products company began experimenting with these butenes especially with an eye towards polymerizing them to iso-octene, which can be hydrogenated to iso-octane.

$$2 \ C_4H_8 \ \text{-->} \ C_8H_{16} \quad Polymerization$$

$$C_8H_{16} + H_2 \ \text{-->} \ C_8H_{18} \quad Hydrogenation$$

There are several processes for conducting the polymerization. The original process used high temperatures, but catalysts were found to make the reaction more efficient. The catalysts were all acids, but different acids and conditions have different advantages. The bottom line is yield and economics. Several pilot plants were built using a thermal polymerization process for turning butene into octene between 1931 and 1934.

Soon, a unified process for manufacturing octane emerged: (1) petroleum could be used to make butene; (2) the butene could be polymerized to octene; and (3) the polymers could be hydrogenated to make octane. By 1934, encouraged by Jimmy Doolittle's arguments that the Air Corps could be convinced to move to a 100-octane aviation gasoline standard, Shell Oil Company (subsidiary of Royal Dutch Shell) built a petroleum-to-octane plant at their East Chicago Refinery. This plant used the Universal Oil Products Company "cold acid" catalyzed polymerization that worked selectively for isobutylene (a.k.a., iso-butene) yielding high-purity iso-octene, which was hydrogenated to 100-octane iso-octane (2,2,4-trimethylpentane). Shell Oil quickly introduced plants for using "hot acid," which polymerized both iso- and n-butene to octene. It was upon this process that Shell hoped to supply the needs of the Army Air Corps; but when the Army Air Corps accepted 100-octane as their standard in 1938, this new market initiated an octane competition between the major oil companies. The search for a better way to make octane was begun and Shell would soon find its processes obsolete.

Before it was superseded, the Shell Oil process spread through the industry in the U.S. and overseas. It was still the method of choice in Britain when the war started in 1939 and through the Battle of Britain in 1940. However, in 1941, Imperial Chemical Industries (ICI) of England applied the Houdry dehydrogenation process to convert butane to butene.

Dehydrogenation

C_4H_{10} butane → C_4H_8 butene + H_2

This allowed them to obtain more octane from a barrel of oil. The hydrogen produced by dehydrogenation was used to hydrogenate the octene polymer.

Also, in 1939, Houdry, who was a patriotic Frenchman (wounded in WWI), returned to France and helped the French government adapt his process to the production of high-octane aviation gasoline. Unfortunately, this was not the best time to introduce high technology into the French petroleum industry as they were soon overrun by the Germans. Houdry returned to America where he helped develop the process for making butadiene for synthetic rubber. By 1942, about 90% of the aviation fuel in the world (Axis and Allies) was manufactured from butene/butane obtained by cracking petroleum in the Houdry process. Dehydrogenation of butane to yield additional butene was used in the U.S. and Europe into 1941-42, but it was soon made obsolete by a process we shall describe below.

Meanwhile in Europe (1930-1939)

In the 1930s, the situation is Europe was quite different. While the United States had an essentially unlimited supply of petroleum and the British could obtain petroleum fairly readily from the western hemisphere or Middle East, mainland Europe was starved for petroleum feed stock. The only two countries in Europe[182] that had substantial petroleum reserves were Rumania and Poland. Germany, France, Italy, Austria, Hungary and Czechoslovakia had/have minimal petroleum reserves and were turning to alternate fuels especially ethanol and coal-derived liquid fuels.

In 1935, Germany imported 1.2 million tons of petroleum fuel and Italy imported about a fourth that much. France imported about 1 million tons of fuel in 1935. In Germany, domestic petroleum provided only about 5% of its peacetime needs in 1935. The Germans were also concerned about a short supply of potatoes for ethanol fermentation stock. Rumania was the second largest petroleum producer in Europe. In 1935, Rumania's distilling capacity was 11.8 million tons *per* year and their cracking capacity was 1.5 million tons *per* year; their actual 1935 production was 8.8 million tons. It was known that the octane rating of the straight run gasoline from Poland and Rumania was low (60-61 octane).

---

[182] I'm obviously not counting Russia in this tabulation. There are substantial oil reserves north and east of the Caucasus.

In Britain, aviation gasoline was typically a good straight run fraction fortified with tetraethyl lead. This combination yielded octane ratings up to about 77 to 87, which were the British and German standards. The British did very little cracking until 1938. However, Britain had large stocks of coal and during the process of making coke, a "light oil" is distilled off that contains benzene, toluene, and xylene (BTX).

The Europeans called this *benzole* (in Britain) or *benzol* (in Germany). The British also tended to call gasoline "benzine." The British used their coal-derived *benzole*, which has an octane of about 130, as a blending stock. They would routinely use blends of 10% to 20% *benzole* with straight run gasoline for aviation fuel. One of the problems with going to higher levels of *benzole* was freezing of the benzene and naphthalene (a trace impurity) in the carburetor. These aromatic liquids (fuels/solvents) also destroyed (dissolved) rubber diaphragms in fuel pumps, fuel lines and self-sealing fuel tanks. In 1938, the French also set 70 to 87 octane as their standards for aviation gasolines.

Interestingly, methanol and ethanol were used as the fuels when the British and French set-out with their high compression racing airplanes to establish speed records. The alcohols had high fuel consumption (limiting range) and alcohol also caused corrosion problems in the fuel systems. Ethanol can be considered to be a partially oxidized hydrocarbon and as such about 20% of its fuel value (energy per unit mass) has been lost relative to aliphatic hydrocarbon gasoline.

In Germany, synthetic liquid fuel plants were under construction in 1935 and the Nazi government was predicting that it would be self-sufficient in petroleum within a few years. But by this time, it was established that the octane of most of this synthetic gasoline, obtained indirectly from coal, would only be about 47, even lower than straight run distillates. Moreover, it contained enough olefins to cause a "gum" problem. If the synthetic fuels were hydrogenated to eliminate the gum problem, the octane of the paraffin was only 12. So, the Germans could produce motor fuel for ground vehicles, but aviation fuel was more difficult to produce. The Germans were very interested in cracking technology and were happy to acquire Houdry's process after they overran France.

But the demand for high-octane was an artifact of the Otto-cycle (4-stroke) internal combustion engine. The need for a relatively volatile hydrocarbon was also related to the use of the carburetor associated with the piston engine. Thus, Germany was moving towards fuel-injection and diesel technology even in airplanes[183]. This was never going to be a high-performance weapon. But the Germans had some other ideas about how to get around the "octane-problem."

German engineers had begun developing a new concept for airplanes called the jet engine and early in the 1930s, the government supported this work because it showed promise of solving the octane crisis in Germany. These engines burned essentially kerosene.

---

[183] Diesel engines generally have a poor power to weight ratio. But the Germans build a very high-flying bomber/ reconnaissance aircraft (Junkers Ju-86P) that could cruise at 41,000 feet. The British spitfire had a ceiling of 38,000 feet. The Germans also worked hard at using diesel (2-stroke) engines for airplanes, probably for the same reason. Diesel engines were useful in heavy equipment such as trucks, but required development of fuel injector systems (instead of carburetors). An important result of this work was that German tanks were powered with diesel engines that had less flammable fuel than American tanks that burned gasoline. American tank crews described their vehicles as "Ronsons" after the popular brand of cigarette lighter that used the advertising slogan "lights first time, every time;" and according to German gunners, they did. Fuel injector technology showed up in some of the German aircraft engines. It was especially useful for water-methanol injection used to cool the exhaust gases.

Alkylation of Isobutane (1935-1949)

Standard Oil of New Jersey had carefully followed the efforts of Jimmy Doolittle (1896-1993) and Shell Oil to get the Army Air Corps to adopt the 100-octane standard in 1938. They knew the basic polymerization technology used by Shell (*circa* 1934) and they knew that if they could find a more economical way to accomplish the same transitions from cracked petroleum to octane *via* butane/butene, they would be able to undercut Shell's prices. The method they developed between 1935 and 1941 was called "alkylation."

Alkylation is the direct reaction of isobutane with olefins (especially n-butene and iso-butene) to form octane:

Isomerization

C_4H_{10} butanes → C_4H_{10} isobutane

Alkylation

C_4H_{10} isobutane + C_4H_8 butene → C_8H_{18}

The reaction was similar to the polymerization reaction and was catalyzed by acids, but it eliminated several steps in the reaction sequence (*i.e.*, hydrogenation).

Thus, it was much more efficient and economical than the Shell polymerization process.

Synthetic Gasoline: The Fischer-Tropsch Process

Germany is rich in coal and very poor in petroleum. Everyone in Germany realized its economic and military limitations with respect to liquid fuels in World War I.[184] Between 1920 and 1928, Franz Fischer (1877–1947)[185] and Hans Tropsch (1889–1935) developed a method for synthesizing aliphatic hydrocarbons from coal and water (synthesis gas, CO/H_2), but as long as natural petroleum is available aliphatic hydrocarbons from the Fischer-Tropsch process is inferior (i.e., very low octane) and economically not competitive. But during WWII, Germany did build and operate numerous synthetic fuel plants.

Fischer and Tropisch found that they could obtain liquid hydrocarbons by passing synthesis gas over an iron-alkali catalyst at 400°C and several atmospheres of

---

[184] Germany made critical early moves in WWII to acquire the Polish oil fields and the Romanian oil fields near Ploesti.

[185] No relation to Emil Fischer.

pressure. They called the product "*synthol*" and it contained various oxygen-containing aliphatic compounds. The *Badische* Company took this line of development to make methanol.

$$CO + 2 H_2 \rightarrow CH_3OH$$

It took Fischer and Tropsch more than a decade to discover and develop a catalyst that would convert synthesis gas to petroleum-like hydrocarbons:

$$8 CO + 17 H_2 \rightarrow C_8H_{18} \text{ (paraffins)} + 8 H_2O$$

By the 1926-28 time period, they were using a catalysts of cobalt and copper, which converted synthesis gas to petroleum-like hydrocarbons at about 300°C. They were able to obtain about 130 grams of liquid hydrocarbons (called "*kogasin*") *per* cubic meter of synthesis gas (e.g., about 500 grams of synthesis gas). The typical products were 4% gases (*e.g.*, propane and butane); 62% gasoline-like hydrocarbons; 23% diesel-like hydrocarbons, and 11% waxes. The octane number of the raw gasoline fraction was 47 and it contained about 25% olefins. When this mixture was hydrogenated to eliminate the gum-forming olefins, the octane rating went down to 12 because it contained almost exclusively straight-chain (normal) aliphatic hydrocarbons.

The next step was to take this process out of the laboratory and make it a commercial success. This was done mainly by the *Ruhrchemie A.G.* Company who mastered the chemical engineering problems. Between 1927 and 1932, the Fischer team developed the catalyst and was working to replace the batch process used in the laboratory with a continuous process for commercial manufacture. In the continuous process, the synthesis gas was swept past the catalyst as a moving gas stream and a certain amount of hydrocarbon was formed in each pass through the reaction vessel. The conversion was measurable by the contraction in the volume of the gas (*e.g.*, in the gas phase, a molecule of carbon monoxide occupies as much space as a molecule of octane; so conversion of many small molecules to a few larger molecules resulted in a contraction of the gas). The engineers tried to obtain as much conversion of the synthesis gas to hydrocarbon on each pass as possible; and the total yield could be raised by sending the gas through several stages of reactors (multiple passes). The work continued in 1932-1934 using a Ni-MnO-Al$_2$O$_3$ catalysts on *kieselguhr* (diatomacious earth used as a support). This catalyst produced about 70 grams of *kogasin per* pass at 200°C and 1 atmosphere of pressure. While the chemists worked to improve the catalyst, the engineers were busy solving the heat transfer problem. Namely, the formation of hydrocarbon from carbon monoxide and hydrogen is very exothermic. The catalyst rapidly heated up and either melted the reactor or charred the hydrocarbon to coke.

The chemists continued to develop the catalyst; and between 1933 and 1939, the catalyst that was developed was composed of cobalt-thorium oxide-magnesium oxide on *kieselguhr* (100:5:8:200 by weight). All the German plants between 1938 and 1944 operated using this catalyst and the *Ruhrchemie A.G.* process at 180 to 200 °C and either 1 atmosphere (low pressure) or 10 atmosphere (moderate pressure). The typical yield was 150 grams of hydrocarbon *per* cubic meter synthesis gas (carbon monoxide to hydrogen ratio 1.0 to 2.0) and the 60 to 100 volumes of feed gas were sent through 1 volume of the catalyst *per* hour.

In summary, a synthetic gasoline plant as operated in Germany during World War II included: A conventional water gas generator followed by a compressor. The compressed water gas passed through units to remove hydrogen sulfide and organic sulfides. Part of the water gas was converted into synthesis gas by the water shift reaction. The hydrogen-enriched gas was sent to the first stage of Fisher-Tropsch reactors. After the first stage, high boiling liquids were removed in a condenser and fresh water gas was introduced to make up the pressure. The process continued through one or two additional stages of reactors. The final product was sent to a cooler condenser and activated carbon where most volatile organics were removed. Non-condensable gas (hydrogen and methane) was flared or burned for heat in the plant. The product yields were as follows:

Gasoline	Diesel	Wax
Low Pressure Process (1 atm)		
56%	33%	11%
Moderate Pressure Process (10 atm)		
35%	35%	3%

The plants were sometimes located near *benzol* plants (discussed below) and the *benzol* plants were able to supplement the hydrogen needs of the process in lieu of the water shift unit. During WWII (1939-45), the limitations on synthesis gas production (*e.g.*, hydrogen availability) was actually the limiting factor in the production rate of these plants.

Liquid Fuel Directly from Coal
(The Bergius Process)

During WWII, Germany also employed the Bergius process to produce various organic feedstocks. It was known that coal (with and empirical formula of essentially CH) produced "light oils" consisting almost exclusively of aromatic hydrocarbons (benzene, toluene, and xylene; BTX) when heated without air. These oils were a byproduct of coke (C) formation, which was a key ingredient for conversion of iron to steel. If this reaction is attempted in the presence of oxygen, polynuclear aromatics (PNA) and phenols are obtained (coal-tar creosote). While coal tar creosote is a valuable product and the naphthalene that is produced is a unique route to an important class of compounds, naphthalene and the phenolic mixture do not constitute an attractive liquid fuel. It was discovered that if coal is heated with hydrogen and catalysts, the yield of aromatic hydrocarbons is increased and the yield of coke was suppressed.

The mixture of aromatics produced by this process (*i.e.,* the Bergius process) was called *benzol* (German for benzene) and is an excellent high-octane fuel-blending-stock (*i.e.,* octane about 130). The yield of volatile liquids was about 50% of the weight of coal. After removing the tar acids and bases by washing with aqueous acids and bases, the neutral oil (crude *benzol*) was typically about 60% aromatics (*benzol*), 30% paraffins, and 10% olefins. The finished *benzol* could be blended to enhance the octane of Fischer-Tropsch fuel. Moreover, toluene could be isolated from the mixture and used as the starting material for trinitrotoluene (TNT).

5. Chemical Secrets and Politics during World War II

In the U.S., hostility towards Standard Oil continued long after the trust was broken up (May 15, 1911). With the eruption of WWII following the attack on Pearl Harbor, many U.S. politicians (notably Senator Harry Truman, head of the Senate Special Committee to Investigate the National Defense Program (started March 1941), a.k.a., the Truman Committee) were happy to point fingers and assign blame in the US. In Germany, things were even worse. The German situation became desperate after the defeat in at Stalingrad (1943). And, when the Anglo-Americans achieved air superiority over parts of Germany in mid-1944, the Gestapo was also looking for people to blame with I.G. Farben as the likely suspect. The result was something of an absurd exchange of claims and denials by Standard Oil and I.G. Farben that they had somehow acquired secret technology concerning high octane fuel (both the alkylation technology and the TEL technology). This theme was picked up by revisionist/anti-business historians into the 1970s.

The Secret of Alkylation

Everyone should remember that up to the end of 1941, the United States was officially (and legally) neutral in the war between Britain and Germany. Well into the 1930s, international trade was encouraged by the U.S. government with the countries that became the Axis. It was official U.S. policy to ship gasoline to Japan almost to the time of the attack on Pearl Harbor. After the war started, political opportunists in and out of government intentionally obscured this point. Starting with the Truman Committee hearings in 1942, the major oil companies in the U.S. (*e.g.*, Standard Oil of New Jersey) were persecuted and publicly chastised for trading and cooperating with I.G. Farben. Antony C. Sutton's 1976 book *Wall Street and the Rise of Hitler* is typical of this view.[186] Joseph Borkin (1978)[187] gives a much more balanced discussion of I.G. Farben and Standard Oil in *The Crime and Punishment of I.G. Farben*.

---

[186] Sutton Antony C. 1976. *Wall Street and the Rise of Hitler.* '76 Press, P.O. Box 2686, Seal Beach, California 90740. 220 pp. This is a "revisionist history" and while the facts seem to be accurately presented, the interpretations given by the author are somewhat biased. The primary source for some of the key points made here is hearings before the Committee on Military Affairs 78th Congress, second session, part 16, p. 939 (U.S. Government Printing Office) and the testimony of von Knieriem at the Nuremburg trials. It should be noted that Sutton and other "revisionists" have as their basic premise that there is an international conspiracy of intellectuals and wealthy people to somehow control the world; and they interpret all events in that context. I will not address that premise here.

[187] Borkin, Joseph. 1978. *The Crime and Punishment of I.G. Farben.* The Free Press a Division of Macmillan Publishing Co. New York, NY. 250 pp.

However, in response to the Truman Committee accusations in 1942 alleging Standard Oil's conduct was "treasonous," a Dr. R. T. Haslam (a director of Standard Oil of New Jersey) published an article in *The Petroleum Times* (25 December 1943) entitled "Secrets Turned into Mighty War Weapons Through I.G. Farben Agreement." The spin of the article was that the U.S. had gained important information from the Germans (I.G. Farben) through the various business agreements and working relationships before the war. The article may have exaggerated the point and because some of the work (such as ongoing alkylation research) was classified as "secret," the Haslam article avoided mentioning it altogether.

Understandably, when the Haslam article found its way into Germany (which was now losing the war and was being plastered by American bombers), officials at I.G. Farben were worried about having to explain their alleged duplicity to the Gestapo (whose reputation was even more feared than the Truman Committee). Naturally, I.G. Farben officials formulated their own propaganda to convince their critics (i.e., the Gestapo) that I.G. Farben had actually out-foxed the Americans. (See Borkin, 1978, pp. Chapter 4, *The Marriage of I.G. and Standard Oil under Hitler*, pp. 76-94.) Sutton (1976) published a lengthy excerpt from a memo dated 6 June 1944 (the day of the Normandy invasion) by a Mr. von Knieriem an official of I.G. Farben (Sutton, 1976, pp. 70-72). The first couple of paragraphs ramble about how the Americans invented the octane concept (*circa* 1923) and how this helped the Germans. By 1939, a petroleum chemist or engine designer would have had to be incompetent to not know this history. The memo does contain several interesting points, however, (quoted from Sutton, 1976, pp. 71-72):

> *"Shortly before the war, a new method for the production of isooctane was found in America – alkylation with isomerization as a preliminary step. This process, which Mr. Haslam does not mention at all* [as noted above, the research was now classified in the U.S.], *originates in fact entirely with the Americans and has become known to us in detail in its separate stages through our agreements with them, and is being used very extensively by us."*

As von Knieriem states, the information "...*has become known to us...in its separate stages*...", but definitely not exclusively from Standard Oil. Basically, I.G. Farben was able to take routinely available information and piece together what to do and how to do it. The revisionists historian's thesis that Standard Oil gave secrets to I.G. Farben falls down very quickly and

totally if the contemporary (1935-1940) technical press
is consulted. Probably the most relevant single
document was published in the *Journal of the Institution
of Petroleum Technologists* (volume 24, pp. 303-325) in
1938. This article by S.F. Birch and coworkers of the
Anglo-Iranian Oil Co. was entitled "Saturated high
octane fuels without hydrogenation. The addition of
olefines to isoparaffins in the presence of sulfuric acid."
You cannot be much more specific than that. This
paper starts with a review of the published literature,
which includes reference to the 1936 publication in the
Journal of the America Chemical Society by Ipatiev and
von Grosse. An entire section is devoted to the reaction
of isobutane with isobutene describing reaction
conditions, yields and octane numbers obtained. The
paper ends with a discussion of how to best formulate
100-octane fuels. This was in the open literature of
1938. The Germans undoubtedly knew about
alkylation as it is mentioned in two contemporary
German technical publications (Mader, 1942, and
Jantsch, 1941 and 1943)[188]. And, they cite open

---

[188] Marder, Maximilian. 1942. *Motorkraftstoffe* (motor fuels).
Springer-Verlag, Berlin. This document was published in the
U.S. in 1945 under the Alien Property Custodian Act by J.W.
Edwards of Ann Arbor, Michigan. It tends to be a summary of
pre-war technology, but confirms that the Germans were fully

literature information from the late 1930s for the key insight into the reaction mechanism (*i.e.*, Birch and Dunstan 1939. *Transactions of the Faraday Society* vol. 35, page 1013).

The synthetic hydrocarbons produced by the Germans during the war (Fischer–Tropsch process) were largely straight chain aliphatics. According to Weil and Lane who reviewed German technology after the war (1948, pp. 112-115):

"*The C3-C4 fraction* [of synthetic hydrocarbons from synthesis gas] *generally constitutes about 13 per cent of the crude liquid product obtained from the normal-pressure synthesis and about 7 per cent of that produced in medium-pressure operations. This fraction is of special interest in any consideration of the Synthine process because of the prospects it offers*

---

familiar with the concepts of polymerization and alkylation reactions to make octane from pre-war publications which are cited. There is a section entitled *Isoparaffinische Benzine durch Addition von Olefinen an Paraffine* (iso-paraffin gasoline from addition of olefins to paraffins) pp. 403-418. There is an entire chapter on *Hochleistungskraftstoffe* (high-test fuels) including a discussion of additives to improve octane number. pp. 459-518.

Jantsch, Franz. 1941, 1943. Kraftstoff-Handbuch (Fuel Handbook in German). Published by Franckh's Verlagshanding, Stuttgart, Germany. Franz Jantsch was the Technical Testing Expert for Oppau at Ludwigshafen am Rhein.

for the manufacture of high-octane polymer or alkylate gasoline...."

"It may be of interest to review here some of the uses to which the Germans have put the C3-C4's produced in their commercial processes. In some of the commercial plants, the C3-C4 cut was absorbed in sulfuric acid and hydrolyzed to alcohols, some incidental polymer being also formed and used for blending in motor fuel...."

"The Germans do not appear to have put very strong emphasis on the production of polymer or alkylate from Synthine C3-C4's, although they were somewhat active along this line. Evidently the strong demand for C3-C4 alcohols and the fact that coal hydrogenation was strongly favored (politically) for gasoline production both served to de-emphasize this development. However, research on the 'Iso-Synthesis' [to make high-octane branched-chain aliphatics from synthesis gas] *was continued throughout the war, and polymer plants had been built. The Castrop-Rauxel plant of Klocknerwerke A.G., for example, was found to possess a polymer gasoline unit capable of processing 25 tons of C3-C4 fraction per day at about 2950 pounds per square inch by the phosphoric acid process. On the basis of 10 days of experimental runs, a 45 percent yield of polymer gasoline was expected."*

Overall, polymerization and alkylation appear to have been minor contributors to the German octane pool by the end of the war and most of that must have been based upon the C4-hydrocarbons obtained from refining natural petroleum. *But, what was left to be inspected in 1946, is not necessarily representative of what was in progress in 1942-43.*

Whatever information I.G. Farben received through 1941 was rapidly becoming obsolete as research was shared among oil companies *within* the U.S. extensively during the war and new technologies emerged. Many patent applications were submitted to the U.S. Patent Office and they were all classified as "secret" until 1946. At that time, they were declassified and most of them were published in a single 1948 compendium by the American Chemical Society (Egloff and Hulla, 1948) with this introduction:

> *"A number of oil companies pooled their scientific and technical knowledge on alkylation during World War II, in order to obtain maximum alkylate production from each refinery. This information was secret during the war. It has since been declassified…"*

The original technology utilized sulfuric acid as the catalyst, but this soon gave way to hydrofluoric acid in the US. By 1946, there were 32 alkylation plants operating in the U.S. using sulfuric acid as the catalysts, 27 plants were using hydrofluoric acid catalyst, and one plant was using aluminum chloride. The product of these plants was called "alkylate" and was used as a blending stock. It was about 24% 2,2,4-trimethylpentane. Other isomers (2,3,4- and 2,3,3-trimethylpentanes) added another 25%. By 1945, the U.S. capacity to produce alkylate was 178,000 barrels *per* day.

The standard 100-octane fuel used by the U.S. Army Air Corps during the war consisted of approximately 50% high-octane straight run petroleum fraction blended with isooctane "alkylate" with 3.0 mL tetraethyl-lead *per* gallon.

The Secret of TEL

Tetraethyl lead technology was widely known in the 1930s by the Japanese and Germans. However, Mr. von Knieriem of I.G. Farben stated the following, in his memo of 6 June 1944 that was intended to protect Farben from the investigations by the Gestapo:

> *"Above all,* [Germany benefited from] *improvements of fuels through the addition of*

tetraethyl-lead and the manufacture of this product. It need not be especially mentioned that without tetraethyl-lead the present methods of warfare would be impossible. The fact that since the beginning of the war we could produce tetraethyl-lead is entirely due to the circumstances that shortly before the Americans had presented us with the production plans, complete with their know-how. It was, moreover, the first time that the Americans decided to give a license on this process in a foreign country (besides communication of unprotected secrets) and this only on our urgent request to Standard Oil to fulfill our wish. Contractually we could not demand it, and we found out later that the War Department in Washington gave its permission only after long deliberation."

Sutton (1976) uses this statement to imply the treason of Standard Oil, but, in von Knieriem's own words, "…*the War Department in Washington gave its permission…*" Standard Oil was not violating the law or secretly collaborating with the Nazis. In fact, Sutton (1976) discusses this incident at length. Contrary to von Knieriem's statement, it did not happen *"shortly before [the war];"* the process began in 1934 shortly after Hitler came to power, but before the rearmament of Germany was obvious. It occurred through the Ethyl Gasoline Corporation (formed in 1924 by Standard Oil and General Motors Corporation to manufacture and sell tetraethyl-lead). Because of the work of Jimmy

Doolittle, who both tried to sell high-octane gasoline to the Germans for Shell Oil and argued that the Americans should adopt the 100-octane standard for U.S. military aviation (probably assuming that the Germans would soon do so), the Army Air Corps was more sensitive to the transfer of technology to the Germans in 1934 than was the Commerce Department. Thus, while the U.S. Government allowed the transfer of technology, a minority (but an informed minority) opinion was expressed by the Army Air Corps on 15 December 1934 to the presidents of Ethyl Gasoline and General Motors (quoted from Sutton, 1976, pp. 189-190):

> *"I learned through our Organic Chemicals Division today (15 December 1934) that the Ethyl Gasoline Corporation has in mind forming a German company with the I.G. to manufacture Ethyl lead in that country.*

> *"I have just had two weeks in Washington, no inconsiderable part of which was devoted to criticizing the interchange with foreign companies of chemical knowledge which might have military value. ...*

> *"It should seem, on the face of it, that the quantity of Ethyl lead used for commercial purposes in Germany would be too small to go after. It has been claimed that Germany is secretly arming. Ethyl lead would doubtless be a valuable aid to military aeroplanes.*

"I am writing you this to say that in my opinion under no conditions should you or the Board of Directors of the Ethyl Gasoline Corporation disclose any secrets or 'know how' in connection with the manufacture of tetraethyl lead to Germany.

"I am informed that you will be advised through the Dyestuffs Division of the necessity of disclosing the information which you have received from Germany to appropriate War Department officials."

Although this letter is perceptive, so was Billy Mitchell. At the time that the letter was written (1934), the best the Air Corps could say was *"It has been claimed that Germany is secretly arming."* So if the U.S. Government was not willing to take a position on the rearmament of Germany, why should a private corporation be expected to act as if the U.S. were at war? Hindsight is always 20/20. Nonetheless, on 12 January 1935, E. W. Webb (the President of Ethyl Gasoline Corporation) responded to the Army Air Corps offering to insert a clause in the contract to guard against technology transfer. Ultimately, Ethyl G.m.b.H. was formed to manufacture tetraethyl-lead in Germany and a firm was also formed in Italy. Nonetheless, Germany was still not self-sufficient in tetraethyl-lead, as Germany imported 500 tons of TEL from the U.S. in 1938. Although this may sound like a lot, the U.S. production in 1941 was 50,000 tons and in 1944 the U.S. produced 100,000 tons of TEL.

Moreover, the limiting component of the anti-knock liquid (at least in the U.S. and the U.K.) tended to be the ethylene dibromide (not TEL). There was not adequate supply of bromine from brine wells at the time and a large plant was built by the Ethyl-Dow Chemical Company at Cape Fear, North Carolina to capture bromine from sea water.

After World War II

There is nothing magic about isooctane (2,2,4-trimentylpentane). It is only one of many highly branched isoalkanes with a molecular weight (volatility) convenient for use as internal combustion fuel. As a matter of fact, among the isoalkanes, isooctane has one of the lower octane-numbers and is one of the least susceptible to octane enhancement with tetraethyl lead. It was recognized as early as 1926 that 2,2,3-trimethylbutane (a.k.a., triptane) had a higher unleaded octane than isooctane (*i.e.*, 100 at lean mixture and 165 at rich mixture). Through 1938, no more than gallon-size quantities of triptane had been synthesized for research. In 1941, the Army tested triptane in the Pratt & Whitney R-1830 engine and obtained about

1,400 hp (Schlaifer and Heron 1950)[189] as compared with 1,200 hp with 91/96 octane grade. Moreover, with 3 mL of TEL *per* gallon, the octane was increased by about 45 octane numbers.

At that time, the synthesis of triptane (by Dow Chemical) was *via* an organo-magnesium alkylation, which would have consumed vast quantities of magnesium metal (2 pounds of magnesium metal *per* gallon of fuel) and cost about $30/gal. The General Motors Company conducted further research and built a triptane plant that could produce about 150 gallons *per* day in 1943 without using magnesium. This plant provided testing samples until 1945. According to Schaifer and Heron (1950, p. 657), "*…very few laboratory engines used for fuel evaluation with supercharging were strong enough to permit triptane + 4 cc lead to be appraised.*" That is, they did not have compression ratios high enough to fully test the fuel. Thus, tests were done with blends of triptane and standard 100 (lean)/ 130 (rich) octane fuel. Such a blend produced 2,800 hp in an Allison V-1710 engine. Without TEL, triptane has little advantage over isooctane as a blending stock. With introduction of the turbojet

---

[189] Schlaifer, R. and Heron, S.D. 1950. Development of Aircraft Engines and Development of Aviation Fuels (Two Studies of Relations between Government and Business). Harvard University, Boston, MA.

engine (which burns kerosene), the era of cost-is-no-object high-octane aviation fuels was over and triptane was never developed commercially.

In 1949, Universal Oil Products Co. announced the "platforming" process in which a platinum catalyst converted aliphatic naphtha into high-octane aromatics (a mixture of benzene, toluene, ethylbenzene and xylene called BTEX) by reforming and dehydrogenation. The product was called "platformate" and became the basis of eliminating TEL from civilian (automotive) use in the 1980s.

6. Organometallic Reagents and Poly-olefins

Karl von Auwers (1863–1939) received his PhD in 1885 from August von Hofmann in Berlin. After that, he worked with Victor Meyer, and on the verge of WWI in 1913, he became the head of the chemistry department at University of Marburg in the Hesse district. Karl Ziegler (1898–1973) was a local student who had excelled in his primary education and entered Marburg in about 1915. Meanwhile, Georg Wittig (1897–1987) had begun studies at the University of Tübingen in 1916. In 1918, both Ziegler and Wittig were drafted into the German army and served at the front. Wittig was captured by the British and was a prisoner of war until 1919. While Zeigler was able to return to school with little difficulty, Wittig found access to Tübingen impossible and was near dropping out of science until Auwers opened doors for him at Marburg. Ziegler (Ph.D. 1920) and Wittig (Ph.D. 1923) were both students of Auwers and both studied for their habilitation up until 1926. Hans Meerwein followed von Auwers as head of the department in 1926 and accepted Wittig as a lecturer. Meanwhile, Ziegler went to the University of Heidelberg. Thus, these two good friends parted ways.

Karl Zeigler: The Early Years

Ziegler's career was very successful and led him through a series of important and lucrative positions, but this was not without controversy. In 1934, he was denounced by Heidelberg Nazi Party officials for associating with Jews and in 1936 his opportunity for advancement was directed into research rather than management because he was regarded as politically unreliable. In 1935/36 he was sent to the University of Chicago on sabbatical. Apparently, soon after his return, he was induced to become a "patron member" (i.e., he provided funding) of the Schutzstaffel (SS). This presumably opened his options and in 1943,

Ziegler followed Frans Fischer as Director of the Kaiser Wilhelm Institute for Coal Research. Although Ziegler had little interest in coal research as applied to liquid fuels, he was given freedom to look for useful applications of coal-derived feedstocks. After WWII, he focused on polymers for plastics. Now let's look at his technical achievements through these years.

His initial interest was in free radicals, which had been prepared while trying to couple triphenylmethyl chloride with alkali metals by Moses Gomberg in 1900. Working with other compounds, Ziegler discovered that potassium will cleave ethers directly (1923-24). In particular, potassium reacts with methyl 2-phenylisopropyl ether to produce potassium methoxide and 2-phenylisopropyl potassium which is bright red.[190] He also observed that phenylisoproply potassium (red) in diethyl ether reacted with 1,2-diphenylethene (a.k.a., stilbene) to yield a yellow product (1928).[191] This was the first indication that free radicles or anions might add to an olefin. Meanwhile in Russia in 1928, Sergei

---

[190] K. Ziegler and F. Thielmann. *Ber.* 56:1740 (1923).

[191] K. Ziegler and Bahr. *Ber.* 61:253 (1928).

Vasiljevich Lebedev had been trying to make synthetic rubber with some success since 1910. Perhaps inspired by Ziegler's observations, he developed a method for polymerization of 1,3-butadiene with sodium in 1928 (see above).

Zeigler also experimented with lithium. Reaction of lithium with, e.g., n-butyl chloride (e.g., 35°C in benzol) produced butyl lithium, which was similar to the Grignard reagents in reactivity (1930).[192] The technique of making "lithium sand" by stirring melted lithium rapidly while cooling in a high boiling aliphatic solvent was introduced some years later.[193]

Zeigler and Natta:
Alkylaluminum compounds and Polyolefins

Ziegler's work from the early 1930s through the end of WWII is fairly obscure. Using the metal alkyls as coupling agents, he succeeded in closing large rings by reactions in high dilution in the late 1930s.

---

[192] K. Ziegler and H. Colonius. *Liebigs Ann. Chem.* 479:135 (1930).

[193] T.D. Perrine and H. Rapoport. Preparation of Organolithium Reagents. *Anal. Chem.* 20(7):635–636 (1948).

The technique was used to close the 6-membered ring during synthesis of cantharidin in 1942.[194] The synthesis was improved by others in 1951.[195]

Drawing by Edgar181, source Wikimedia Commons.

According to his Nobel lecture (1963), he succeeded in reacting lithium alkyls with 1,3-butadiene and ethylene during this period (1943-45). This would be consistent with the German research on synthetic hydrocarbons, but linear hydrocarbons have such low octane ratings that the work was likely not followed up aggressively.

In the post-war years, Zeigler followed up the idea of alkylation of olefins (see above discussion concerning manufacture of isooctane fuels) with the objective of forming longer linear polymers. Rather than using Bronsted acids, he chose Lewis acids in particular triethylaluminum. Using this technique, Ziegler and

[194] Ziegler used N-bromosuccinimide (introduced by Alfred Wohl (1863-1939)) to chlorinate methyl ketone groups, which were then coupled with alkali metal.

[195] G. Stork et al. Cantharidin. A stereospecific total synthesis. *J. Am. Chem. Soc.* 73(9):4501–4501 (1951).

Hans-Georg Gellert managed to build polymers with ethylene to about 100 monomer units.[196] These products were useful because they decomposed to the 1-olefin and aluminum hydride at high temperature and the terminal olefins could be used to alkylate benzene to make biodegradable detergents (e.g., C_{18}-alkylbenzenesulfonic acid salts).[197]

In the 1930s, Ziegler had attempted to sublime lithium alkyls, but found that they decomposed to lithium hydride. Attempts to reverse this reaction, e.g., by adding lithium hydride to olefins under pressure were doomed to failure by the high lattice energy of lithium hydride. However, once Schlesinger *et al.* produced lithium aluminum hydride in 1947 (see below) and found it was soluble in ether, the problem of lattice energy was resolved. Thus, in 1948 Ziegler and Gellert found that lithium aluminum hydride reacted smoothly with ethylene. At 100°C (1 atm), lithium

---

[196] K. Ziegler and H.-G. Gellert, *Ann. Chem.*, 567:195 (1950).

[197] Previously branched chain alkylbenzenes had been used and these are not readily biodegraded. Thus, the detergents produced foam in industrial waste water released into streams.

tetraethylaluminate could be isolated; but at 180-200°C and 100 atm long polymers of ethylene are formed.[198]

It was found that direct reaction of aluminum with hydrogen was not effective (unlike LiH, AlH_3 has a low lattice energy); but with H_2 and an olefin, the corresponding aluminum alkyls are formed.

It was also observed that in some cases, the addition reaction was prematurely terminated. Examination of the materials indicated that traces of nickel in the aluminum produced the effect (Raney nickel). While trying to eliminate this effect, Ziegler et al. found that some transition metals ($TiCl_4$) delayed the elimination of the terminal olefin allowing truly long polymer chains to form. Thus, mixed-metal catalysts were invented.

Giulio Natta (1903–1979) an organic chemist in Italy who (like-Ziegler) rode out WWII in research, became familiar with Ziegler's work in 1952. By 1954, he was aware that the poly-olefin polymers occurred in different isomeric forms (isotactic, syndiotactic, and

---

[198] K. Ziegler and H.-G. Gellert. *Brennstoffchemie.* 33:193-200 (1952).

atactic) with different physical properties. These structures were confirmed by x-ray crystallography. Natta's work not only helped understand the physical properties of poly vinyl compounds, it assisted in understanding the mechanism of the polymerization reaction. He received the Nobel Prize with Ziegler in 1963.

Ziegler-Natta Polymerization Mechanism
Author: Stuart Prescott and Robert G. Gilbert; Source:
http://discovery.kcpc.usyd.edu.au/9.2.1-short/9.2.1_Polyethene2.html

Wittig and Gilman: Lithium Exchange Reactions
Like Ziegler, Wittig's career was influenced by the political changes in Germany between 1930 and 1945. As noted above, Wittig remained at Marberg until 1932 when he moved to the Braunschweig University of

Technology. During this time, his interest were very similar to Ziegler; specifically, he was interested in forming free radicals, especially phenyl-stabilized di-radicals (i.e., diyls) produced by cleavage of strained rings with vicinal phenyl groups. These were being prepared in several steps by reacting phenyl Grignard with corresponding dicarboxylic acid esters. Wittig found that the phenyl Grignard was ineffective and used the methods of Ziegler and Colonius[199] to prepare phenyl lithium from bromobenzene (1930), which worked.[200] However, the compounds he formed did not show the expected tendency to produce diradicales. Thus, he tried to make the corresponding anisole compounds, which were expected to favor the radicals.

---

[199] K. Ziegler and H. Colonius. *Liebigs Ann. Chem.* 479:135 (1930).

[200] G. Wittig. FROM DIYLS TO YLIDES TO MY IDYLL
Nobel Lecture, 8 December 1979.

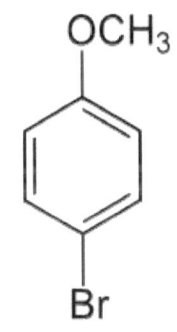

OCH₃

Br

Drawing by
Yikrazuul, source
Wikimedia Commons

Instead of obtaining the expected p-lithioanisole from reaction of lithium with p-bromoanisole, he found that the lithium exchanges with a hydride to produce more stable 2-methoxy-5-bromophenyl lithium and anisole. He had discovered lithium-hydride exchange.

At this point, Wittig's interest shifted more towards the reactivity of phenyl lithium. In the meantime (1932), organolithium methodology was improving rapidly and Henry Gilman (1893-1986) came into the picture.[201]

Nazi politics progressively made life more unpleasant and in 1937 Wittig moved to the University of Freiburg where Hermann Staudinger was the department head. This move was partially facilitated because Wittig had

---

[201] H. Gilman, E.A. Zoellner and W.M. Selby. An improved procedure for the synthesis of organolithium compounds. *J. Am. Chem. Soc.* 54(5):1957–1962 (1932).

supported the polymer hypothesis of Staudinger in his book on stereochemistry.

Wittig discovered that lithium exchanged with iodide, bromide and even chloride (but not fluoride) on aromatics compounds in equilibria that resulted in the most stable aryl-lithium compound.[202] Gillman made the same discovery almost simultaneously using alkyl (e.g., methyl, ethyl, propyl and butyl) lithium reagents.[203] It is said that Gilman and Wittig reached "a gentleman's agreement" in which Gilman worked with alkyl lithium reagents and Wittig worked with phenyl lithium.[204]

---

[202] 6. G. Wittig and U. Pockels. *Ber. Dtsch. Chem.* Ges. 72:89 (1939).

[203] H. Gilman and F.W. Moore. Some factors affecting halogen-metal interconversions. *J. Amer. Chem. Soc.* 62(1):1843-1846 (1940).

H. Gilman et al. Some interconversion reactions of organolithium compounds. *J. Amer. Chem. Soc.* 62(9):2327-2335 (1940).

[204] G. Wittig and G. Fuhrmann. Ber. Dtsch. Chem. Ges. 1940(73):1197.

Wittig's most important work was done between 1937 and 1956. In the case of fluorobenzene, lithium exchanged with the hydride adjacent to the fluorine (1942). Then the unexpected happened, lithium fluoride (which has a favorable lattice energy) split out of the benzene yielding very reactive dehydrobenzene (C_6H_4)[205] which reacts as a dieneophile.[206]

The Wittig Reagents and Reactions

In the late 1920 and 1930s, Thomas Stevens (1900-2000) had discovered the rearrangement of the benzyl group (phenylmethylene-) from nitrogen of an ammonium salt to an adjacent carbon atom that had been deprotonated by a base.   Marcel Sommelet (1877-1952) and Charles Roy Hauser (1900-1970) observed that when the benzyl protons are the most acetic attached to an ammonium (e.g., in trimethylbenzylammonium), the Stevens rearrangement is accompanied by an unexpected de-protonation of one of the methyl groups on the nitrogen and attack of that nucleophile on the ortho-position of the benzyl with displacement of the

---

[205] G. Wittig, *Naturwissenschafen* 30:696 (1942).

[206] G. Wittig and L. Pohmer, *Chem. Ber.* 89:1334 (1956).

dimethylamine from the benzylmethylene. The resulting non-aromatic intermediate immediately undergoes a hydride shift (from the ortho position of the ring to the methylene group) to produce a 1-dimethylaminomethyl-2-methylbenzene. This is known as the Sommelet-Hauser rearrangement and these mechanisms have been hotly debated.

Wittig apparently became interested in these observations and moved on to experiments trying to abstract protons from tetramethylammonium chloride, which yielded trimethylammonium methylide. Wittig found that phosphonium salts were even more easily formed because the phosphorus could essentially form a double bond with the methylide (10 electrons around P).

Drawing by Smokefoot, source Wikimedia Commons

When attempting to attach the ylide to benzophenone, Wittig discovered to his amazement that the oxygen was replaced with the methylene group.[207] The reaction proceeds through a four-membered ring (which can be isolated in some cases) and is driven to the final product by the stability of phosphine oxides (R_3PO). These "Wittig reagents" found almost immediate commercial application (e.g., in the synthesis of vitamin A):

Drawing by NEUROtiker, source Wikimedia Commons.

7. Polyesters and Polyamides (Nylon)

After his success with neoprene at DuPont in 1931, Wallace Carothers moved on to new problems. He developed processes for making polyesters from dicarboxylic acids and glycols. In 1935, he hit on a way to manufacture a polyamide polymer by

---

[207] G. Wittig and G. Geissler, *Justus Liebigs Ann. Chem.* 580:44 (1953).

condensation of a diamine (hexamethylenediamine) and a dicarboxylic acid (adipic acid).

$$n \; \underset{HO}{\overset{O}{\underset{\|}{C}}}{-}R{-}\underset{OH}{\overset{O}{\underset{\|}{C}}} + n \; H_2N{-}R'{-}NH_2 \longrightarrow \left[\underset{}{\overset{O}{\underset{\|}{C}}}{-}R{-}\underset{}{\overset{O}{\underset{\|}{C}}}{-}\underset{H}{\overset{}{N}}{-}R'{-}\underset{H}{\overset{}{N}}\right]_n + 2 \; H_2O$$

Drawing by Calvero, source Wikimedia Commons

The name "nylon-66" was eventually applied, and as a thermoplastic, it could be melted and extruded into fine fibers. It first entered commerce as toothbrush bristles in 1938.[208] By 1940, the extrusion technology had improved and it appeared as nylon stockings, which were an immediate success as a cheap, durable and stylish replacement for silk.

The potential of nylon was recognized by I.G. Farben and they had to find a way around the DuPont patents. Thus, in 1938 Paul Schlack (1897–1987) developed a method for making "nylon-6" via caprolactam:

Drawing by Nuklear, source Wikimedia Commons

---

[208] Carothers had suffered from serious depression for years and shortly after the death of his sister, he committed suicide with cyanide April 29, 1937. The commercialization of nylon was completed by several of his colleagues.

You may recall that Ernst Beckmann had already provided a convenient method for synthesis of caprolactam from cyclohexanone (Beckmann rearrangement) in 1886.

8. Infrared Spectroscopy

In 1800 William Herschel (1738-1822) discovered that there were electromagnetic rays that carried heat beyond the red end of the visible spectrum, i.e., the infrared. It was not until the 1870s that heat was accepted to be kinetic energy of moving molecules (primarily "vibrations"). After the development of statistical thermodynamics by Boltzmann (1890), quantum theory by Planck (1900), and the explanation of atomic spectra by Bohr (1920), it became likely that similar spectra must be associated with heat (i.e., infrared radiation). But, the path was not that direct. William Coblentz (1873–1962) entered Cornell University as a graduate student in 1900. He constructed a primitive spectrometer with a glass prism and painstakingly calibrated wavelengths and intensities of light from sources before and after passing through samples. In this way, he point-by-point created spectra of various substances. From these data, he was able to identify characteristic frequencies of various functional groups.

In the 1920s, the possibility of emissions of absorbed infrared radiation was raised and proven (Raman effect). This technique is useful for assigning certain vibrations to specific emissions/absorptions, but is not particularly popular for general work.

The first fully automated IR spectrometer was built at BASF (Germany) in 1937. During WWII (1939-1945) the US used alkylation of olefins to produce high octane gasoline (100 octane) and all this technology was kept secret.[209] IR spectroscopy was used extensively to characterize hydrocarbon molecules. These data were made public by the American Petroleum Research Program in 1946. The first scientific papers from academia on application of infrared spectroscopy in organic chemistry date from 1947, but within months the number of papers exploded (see for example the pubmed.gov database) as IR spectroscopy and instrumentation spread.

---

[209] For example: M. H. Gorin , C. S. Kuhn Jr., C. B. Miles Mechanism of Catalyzed Alkylation of Isobutane with Olefins. *Ind. Eng. Chem.*, 38 (8): 795–799 (1946).

In principle, molecular vibrations involve all the atoms of a molecule. The molecule does not "know" it is held together by chemical bonds; the molecule is a collection of masses held in place by a collection of force constants. Nonetheless, certain vibrations of complex molecules can be characterized as stretching or bending modes (i.e., degrees of freedom). Thus, functional groups can often be identified by characteristic absorptions (relatively independent of the overall structure of the molecule). The IR spectrum conventionally is plotted as transmission: 4000 cm^{-1} (wave numbers) to about 200 cm^{-1}. The lower frequency cut-off is generally determined by the sample holder (e.g., salt plates (NaCl, KBr, and CsI) have been extensively used to hold the sample in the beam). In recent years, Fourier transform technology has made the time required to obtain IR spectra much shorter with improved sensitivity and reflectance techniques simplify sample positioning.

There are three points that should be remembered about IR spectra: (i) the time scale for IR is literally the frequency at which molecules vibrate ($\sim 10^{14}$/s) and anything that happens more slowly (e.g., rotations and conformational changes) will be observed as separate species by IR. Thus, molecules such as diethyl ether that exist in numerous conformations in solution will be observed as a collection of different conformations (e.g., broad absorptions associated with the structural conformations) in the liquid form; but when coordinated to a metal in a solid phase where one conformation is "frozen" in place, these structural absorptions will be sharp representing a unique conformation. (ii) Because the energy level spacing of vibrations is relatively large, most vibrational degrees of freedom molecules will only be populated in the ground state at ambient temperature. In addition, the zero-point vibrational energy will be very different for isotopes of different mass (H versus D "deuterium") leading to large shifts in the absorptions and isotope effects in kinetics (rates) and thermodynamics (equilibria). (iii) for small molecules where a vibration (e.g., bond stretching) must change the rotational moment of inertia, there must be significant coupling of the rotational and vibrational transitions. This gives rise to vibrational spectra, which are split into numerous components. The intensity of the components is determined by the Boltzmann distribution of molecules in the rotational energy levels.

9. Synthetic Drugs and Pesticides

Arsenical drugs had been developed in the early 1900. This application of the principles of toxicology and pharmacology would continue with development of new drugs and pesticides. The need and market for drugs, especially those that could cure bacterial infections, was obvious; and by the 1930s, companies that had been doing chemical research for decades had large libraries of potential drug candidates.

The Sulfa Drugs

In the 1930s, the Bayer Company in Germany[210] was using its resources to sort through the numerous compounds that had been synthesized during the search for economically valuable dyes. But now they were looking for bioactive compounds. In 1932 Gerhard Domagk (1895–1964) found that a red dye synthesized in 1909 was active against bacteria. Clinical trials followed (1932-34) and the compound

---

[210] An ability to read German was considered essential for organic chemistry PhDs as late as 1970.

(Prontosil) was introduced commercially in 1935 as an antibiotic. It worked remarkably well.

Prontosil
Drawing by Edgar181, source Wikimedia Commons

Sulfanilamide
Drawing by Choij, source Wikimedia Commons

Also, in 1935, scientists in France discovered that Prontosil is rapidly metabolized to sulfanilamide, which is the actual biologically active compound. Prontosil was widely accepted, used liberally, and did much good. However, it led to a tragedy and a new law.

Most sulfanilamide drugs were sold in tablet or capsule form. But, in 1937, a company decided to take advantage of the red color of Prontosil and produced an "elixir" of the compound in what they though was a harmless solvent (72% diethylene glycol, and 16% water) with raspberry flavoring. Ironically, the term "elixir" was legally reserved for alcohol (i.e., ethanol)-

based drugs. This may have induced some people to view it as an alcoholic beverage.

At the time, few people realized that diethylene glycol was toxic. About 240 gallons of this material was distributed in pint bottles into a market targeted at children. Parents bought it and children loved it; and within a month half-a-dozen people were dead of kidney failure. The toxicity is due to metabolism to 2-hydroxyethoxyacetic acid, which seems to be re-absorbed in the renal circulation causing excess acidity (metabolic acidosis). The Food and Drug Administration (FDA) used its limited authority covering *mis-branding* to spearhead the removal of the improperly-labeled "elixir" from the market. This event, motivated Congress to pass the 1938 Food, Drug, and Cosmetics Act, which gave FDA power to require *safety* testing[211] of drugs. The actual active ingredient was not to blame for the toxicity and a large family of "sulfa drugs" (antibiotics) were developed in the 1930s.

---

[211] However, authority of FDA to require *efficacy* testing did not come until 1962 (Kefauver-Harris Drug Amendments).

The Quinoline Drugs

Meanwhile, malaria continued to be the bane of life in the tropics. In the Congo basin, it is estimated that nearly 50% of infants died of malaria before they reached puberty (and they still do). Of course, this was a secondary problem for Europeans who wished to colonize and exploit tropical resources. Belgium, in particular, with its long history in the Congo and Germany with its history in Cameroon were interested in new drugs against malaria. The British had more or less standardized "Gin and Tonic (i.e., quinine water)" as the routine for its colonial officials in India and Africa. And, once seeds were smuggled out of Peru by the Dutch, 97% of the quinine was produced on plantations in Java in the 1930s.

The research power of Bayer chemists and physicians was also working on the malaria problem. The structure of quinine was deduced in 1908 by Paul Rabe, who also achieved a partial synthesis in 1918 by closing the quinuclidine on the side chain.[212] Thus, research

---

[212] P. Rabe P and K./ Kindler. *Ber. Dtsch. Chem. Ges.* 51:466–467 (1918).

centered on synthesizing quinoline compounds with simpler side chains.[213] The first drug they discovered was known as Pamaquine (1926). The full synthesis begins with anisole, which is nitrated, reduced to p-methoxyaniline and N-acetylated. A second nitration (ortho to the amide) followed by hydrolysis of the amide produces 2-nitro-4-methoxyaniline. The Skraup reaction (1880, see above) forms the pyridine ring. The nitro group is reduced (Sn/HCl) and the side chain is added.

Pamaquine (1926)   **Chloroquine (1934)**

Drawing by Fvasconcellos, source Wikimedia Commons

Pamaquine was found to be effective in animal studies and small human trials conducted in Leopoldville by

[213] Methylene blue, which Ehrlich had found to have some antimalarial effects had some of these features.

the Belgian clinic (1927). It was then tested by the British in India and was the leading synthetic drug used into the early part of WWII (1942). When the Japanese occupied the Dutch East Indies (including Java) at the start of WWII (1941), they controlled all the natural quinine. Thus, the U.S. Marines sent into the islands of the South Pacific in 1942, used pamaquine and/or a similar drug quinacrine (Atabrine™). However, these were judged to have too many side effects. In response, the U.S. began a massive set of trials looking for anti-malarial drugs.

In 1934, the Bayer laboratories (Hans Andersag, 1902-1955) had produced a quinoline compound called "Resochin," which was effective against malaria, but which was judged to be too toxic for general use. This information was available to the U.S. Winthrop Company because of its affiliation with Bayer. When the Allies invaded North Africa (1943), they captured French soldiers who were using the German product Sotoquine™, which was known to be less toxic (but less effective) than Resochin. Thus, Resochin (now known as chloroquine) was included in U.S. trials and proved to be both safe and effective by 1944. Chloroquine has continued in use since that time, but its effectiveness has declined (especially after 1970) as the protozoa has developed resistance.

Dichlorodiphenyltrichloroethane

Reaction of chlorobenzene with chloral (trichloroacetaldehyde) produces a mixture of products known as DDT.

p,p′-DDT

Drawn by Leyo, source Wikimedia Commons

DDT had been synthesized in 1874 (Othmar Zeidler, 1850–1911), but its amazing insecticidal actions were not discovered until 1939 by Paul Müller (1899–1965) working for Geigy in Switzerland. With Europe at war in 1939, Müller was in a neutral country and obtained Swiss, UK and US patents between 1940 and 1943. The U.S. included it in its large war-time trials and after confirming its safety and effectiveness, used it extensively in 1944-45 to kill mosquitoes on Pacific islands by dusting coastal swamps from airplanes as part of the preparation for invasions. It was also used in Europe (e.g., Naples, Italy) to suppress typhus epidemics during and after the war.

The World Health Organization's Attempt to Conquer Malaria

Armed with DDT and chloroquine, the World Health Organization (WHO) set out to eradicate malaria in 1955. This effort was successful in clearing most of Europe, the southern US and some islands of malaria, but it ultimately failed in the Congo basin of Africa. The effort stretched into the 1970s, but was stopped when it was observed that the mosquitoes were become resistant to DDT and the malaria parasite was becoming resistant to chloroquine.

10. Total Synthesis

By the 1940s organic chemists had developed a strong understanding of the structure of organic compounds, their reactivity and many reagents had been invented to execute chemical transformations. But, the exact structure of many natural products that were known to have medicinal or toxicological properties were not know or not confirmed. Techniques such as crystallography, mass spectroscopy and nuclear magnetic resonance spectroscopy (NMR) were either not known or very limited in their capabilities. Thus, the stage was set for total synthesis of complex molecules to become an important technique for determining and confirming structures. This chemistry was dominated for thirty years by Robert Burns Woodward (1917-1979). Woodward and Carl Spaatz (1891-1974) were probably responsible for shifting the epicenter of organic synthesis from Germany and Europe to the United States after WWII. There were soon many chemists taking up the challenge of "total synthesis" with varying degrees of success.

R.B. Woodward: The early years

Many books could be written about the work of Woodward and his many students and post-docs.[214] Here we will only give a hint of the material available. He knew he wanted to study chemistry from a very

---

[214] For example, *Robert Burns Woodward: Architect and Artist in the World of Molecules* edited by Otto Theodor Benfey, Peter John Turnbull Morris. 2001. Chemical Heritage Foundation. Philadelphia, PA. 475 pp.

early age and before entering college had performed may classical experiments. He entered MIT in 1933 and focused on chemistry to the point of neglecting other studies. Nonetheless, he managed to complete his BS in 1936 and PhD in 1937. By 1938, he was at Harvard where he remained for the rest of his life.

To this point, synthetic chemistry had mainly been limited to trial and error making compounds with little stereo control (except for the meticulous work of Emil Fischer) and with little reliance on the principles of physical organic chemistry that had blossomed in the 1930s. Woodward realized that to synthesize biologically active molecules, which the pharmaceutical industry was eager to obtain, would require multiple steps and strict control of stereochemistry.

The molecule that had intrigued chemists for 100 years was quinine. Woodward knew that whoever synthesized quinine from readily available chemicals would have his name in the history books. Reviewing the literature, principally work of Paul Rabe, he realized that Rabe and Kindler had solved one of the most difficult parts of the synthesis in 1918: Conversion of d-quinotoxine into quinine. The synthesis actually yields a mixture of products and isomers, but pure quinine can be isolated from the reaction mixture by selective crystallization with L-tartaric acid from 95% ethanol to obtain di-quinine L-tartaric acid monohydrate salt.

Thus, Woodward targeted synthesis of d-quinotoxine and obtained this compound in 1944. He announced the work as a "total synthesis of quinine."[215] Most people were impressed to the point of ignoring the fact that he had not actually produced quinine. In later years, this synthesis was redefined as a "formal total synthesis" of quinine, which depended on the validity of the work of Rabe and Kindler. Because Rabe and Kindler's work was done with such basic resources and one of the key steps (a reduction with aluminum) was not fully described until much later, their work (and hence Woodward's claim) was viewed with some skepticism until it was repeated with modern methods of analysis in 2008.[216]

---

[215] R.B. Woodward and W.E. Doering. The total synthesis of quinine. *J. Am. Chem. Soc.* 66 (5):849–849 (1944).

[216] A.C. Smith and R.M. Williams. Rabe Rest in Peace: Confirmation of the Rabe-Kindler Conversion of d-Quinotoxine to Quinine. Experimental Affirmation of the Woodward-Doering Formal Total Synthesis of Quinine. *Angew Chem Int Ed Engl.* 47(9): 1736–1740 (2008).

Having established his pattern of work and obtained the necessary reputation to acquire funds and co-workers, Woodward took on progressively more challenging projects. In 1951, Woodward and co-workers announced an approach to the synthesis of steroids and the total synthesis of cholesterol.[217] This was a 40-step process, which firmly established Woodward as the leading synthetic chemist in the world. He was able to achieve stereo-chemical control by carefully planning the steps and, where necessary, introducing protecting and stereo-control elements that could be removed later in the synthesis.

In 1954, Woodward and co-workers announced the synthesis of strychnine through 29 steps yielding 8 mg of product.

Obviously, these syntheses were not valuable as commercial processes because the compounds were generally available from biological sources. *At the time, these efforts were mainly justifiable in terms of confirming the structures of readily available natural products. This justification, however, is somewhat eroded today by modern spectroscopic and analytical techniques.* The many

---

[217] R. B. Woodward, F. Sondheimer, D. Taub. The total synthesis of cholesterol. *J. Am. Chem. Soc.* 73 (7):3548–3548 (1951).

intermediate products isolated and characterized along the pathway, however, sometimes proved to be valuable for production of modified compounds that had enhanced bioactivity.

John Clark Sheehan: Synthetic Penicillin

Alexander Fleming (1881-1955) was the first to observe that a secretion from certain molds (*Penicillium notatum*) inhibited the growth of many bacteria (September 3, 1928). His team was not able to isolate the active ingredient but published their results in June 1929.

Britain was drawn into WWII in 1939 and many of the country's industries were redirected to project to support the military. The value of a powerful antibiotic was not lost on British scientists, generals or politicians. At Oxford University, Howard Florey (1898-1968) and Ernst B. Chain (1906-1979) turned their laboratory into a penicillin factory to grow the mold in quantity for extraction of the active agent. Norman Heatley (1911-2004) and Edward Abraham (1913-1999) extracted and purified the crude penicillin by chromatography on alumina columns.

By 1941, penicillin had been isolated and shown to be practical for use in humans. Because of the war, Florey and Heatley travelled to the U.S. to arrange mass production. This work settled at the Department of Agriculture's Northern Regional Research Laboratory (NRRL) in Peoria, Illinois where culture techniques and improved strains raised the yield many fold.

It was discovered that four U.S. companies had independently begun some work on penicillin and a meeting was held in October 1941 to find a partner for this work, in much the way that petroleum and rubber production would be advance. Little was accomplished until after the bombing of Pearl Harbor (December 7, 1941) and it was clear that the U.S. government would be supporting the research. From these meetings, Merck & Co. took the lead.

Abraham, who was the first to isolate crude penicillin also led the work on determining the structure of penicillin. His experiments led him to the idea of a beta-lactam structure, but the head of his laboratory Robert Robinson and J. W. Cornforth believed it was a less- strained thiazolidine-oxazolone structure (i.e., the two rings are joined by a single bond, not fused).[218]

---

[218] R. Curtis and J. Jones. Robert Robinson and penicillin: an unnoticed document in the saga of its structure. *J. Peptide Sci.* 13(12):769-775 (2007).

Robinson, of course, was very influential and defended his proposal beyond the point of reason in most minds. Indeed, there was still speculation about the structure of penicillin until 1945 when Dorothy Crowfoot Hodgkin (1910-1994) and her team solved its x-ray crystal structure (not published until after the war in 1949)[219], which proved it to be a fused ring beta-lactam.

6-Aminopenicillanic acid

Penicillin core

Drawings by Yikrazuul, source Wikimedia Commons

Meanwhile, John Clark Sheehan (1915–1992) also showed interest in chemistry at an early age, but was much more diverse in his activities than Woodward. He earned his PhD in organic chemistry from the University of Michigan in 1941. As a post-doc, he helped development of the powerful explosive RDX

[219] The long delay in publication is indicative of the fact that everyone who needed to know had already been informed secretly in 1945 and after 1946, the news was of little interest.

before moving to Merck & Co. late in 1941. In early 1942, he was assigned to the penicillin project at Merck.

Production of the drug was ironed out and production increased with the goal of having an excess supply in 1944 (1,663 billion units in 1944) for the anticipated invasions of Europe and Japan. By March 1945, the supply was essentially unlimited for military and civilian use in the U.S. and Britain.

During the war, thousands of chemists at leading pharmaceutical companies worked to determine the structure and develop a synthesis for penicillin. But after the war and the reduction of the price of natural penicillin to a few cents to cure typical infections, most drug companies dropped the work as impractical. In this context, in 1946, Sheehan left Merck and took a much lower salary to be an assistant professor at MIT. He brought the problem of penicillin with him. By 1949, he is believed to have been the only chemist working on synthesis of penicillin. The synthesis was not completed until 1957, but it had a very useful byproduct. It turned out that by the 1950s bacteria were becoming resistant to penicillin. But in 1958, 6-aminopenicillanic acid was isolated by scientists at Beecham and became the starting point for a number of semi-synthetic penicillin derivatives. Moreover, the beta-lactam structure is present in several families of antibiotics.

(1) Penicillin (2) Cephalosporin

Drawings by Fvasconcellos,

source Wikimedia Commons

11. Conformational Analysis

Structural organic chemistry had raced ahead of bond theory from 1870-1930 and it was generally understood that rotation was restricted around double bonds and rotation was free around single bonds. But, it was only after a theoretical understanding of molecular orbitals and hybridization was available (1930-40) that chemists came to fully appreciate the details of structural theory.[220] Perhaps the first hints of more detail were noted when it was found that the entropy of ethane was less than expected for completely free rotation about the C-C single bond. While it was not known whether the barrier was based on attractive forces or repulsive forces, it was obvious it should be universal in aliphatic compounds. It became apparent that the barrier was *repulsive* when cyclohexane was considered. The preference for the chair conformation over the boat conformation of cyclohexane (both free of ring-strain but representing alternatives of staggered

---

[220] D.H.R. Barton. The principles of conformational analysis. Nobel Lecture, December 11, 1969.

and eclipsed conformation of hydrogens) was established by Hassel and Ottar (1947)[221]: The diagram below show relative energies of various conformations.

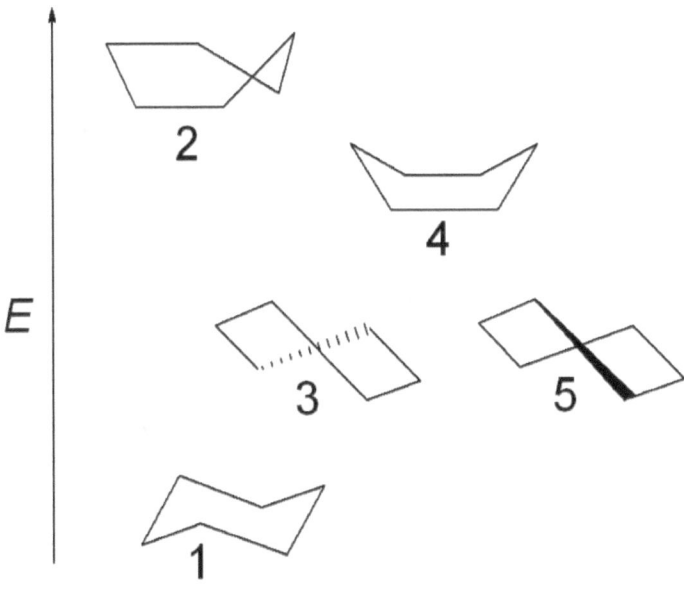

Drawing by Cvf-ps, Source Wikimedia Commons.

Modern terminology was introduced in 1954.[222] It was soon recognized that the most stable conformations

---

[221] O. Hassel and B. Ottar. *Acta Chem. Scan.* I:929 (1947).

[222] D.H.R. Barton, O. Hassel, K.S. Pitzer and V. Prelog. Nomenclature of cyclohexane bonds. *Science.* 119(3079):49 (1954). The same letter was published a few days earlier in *Nature.*

place bulky substituents (e.g., t-butyl) into equatorial positions where there are farthest apart.[223] In decalin (two fused cyclohexane rings) both the cis- and trans-fusions can achieve the chair conformation in both rings with quite different disposition of the rings:

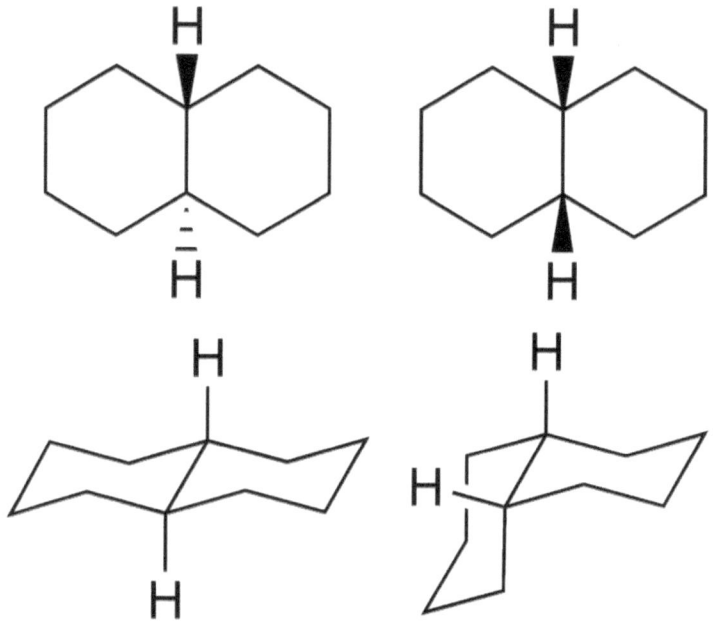

Drawing by Azazell0, source Wikimedia Commons.

[223] C.W. Beckett, K.S. Pitzer, R. Spitzer. The thermodynamic properties and molecular structure of cyclohexane, methylcyclohexane, ethylcyclohexane and the seven dimethylcyclohexanes. *J. Am. Chem. Soc.* 69 (10):2488–2495 (1947).

Derek Barton (1918–1998), who was familiar with the work of Odd Hassel on cyclohexane, was a visiting professor of natural products at Harvard (1949-1950). By 1950, the medical importance of steroids was known and the Woodward team at Harvard was in the process of synthesizing cortisone.[224] This interaction probably prompted Barton to assimilate his knowledge of cyclohexane rings and the relative rates of reaction of various substituents on steroid rings. In 1950, he described how steroid rings would form and the relative reactivity of substituents in a paper of only 4 pages.[225] This paper founded the subject we now know as conformational analysis.

Conformational analysis resolved some of the mystery associated with elimination reactions, which had been brought into focus as the conflict between Hofmann's

---

[224] R.B. Woodward, F. Sondheimer, D. Taub. The total synthesis of cortisone. *J. Am. Chem. Soc.* 73(8):4057–4057 (1951).

[225] D.H.R. Barton. Conformation of the steroid nucleus. *Experientia.* 6:316-319 (1950).

rule[226] and Saytzeff's rule (Zaitsev's Rule).[227] The Ingold-Hughes group had published a long series of papers on elimination reactions ending in 1948.[228] They resolved the differences between E1 and E2 eliminations and found that E2 elimination was facilitated by anti-periplanar arrangement of the leaving groups. This often becomes the controlling factor in some eliminations in cyclohexane rings. However, in a few cases, syn-periplanar elimination occurs when anti-periplanar orientation is not possible (e.g., in a norbornene system):

Rate of E2 Elimination

anti-periplanar >> syn-periplanar >> gauche

---

[226] When the leaving group is very large (e.g., trimethylamine) the less substituted olefin will be favored.

[227] The most substituted (most stable) olefin is favored. M.L. Dhar, E.D. Hughes, C.K. Ingold et al. Elimination reactions in organic chemistry. *Nature*. 147:812-813 (1941)

[228] M.L. Dhar, E.D. Hughes, C.K. Ingold et al. Mechanism of elimination reactions. Part XVI. Constitutional influences in elimination. A general discussion. *J. Chem. Soc.* 1948:2093-2119.

12. Metal Hydride Reducing Agents

Elemental boron was produced in small impure quantities from borax or boric acid by various chemists in the early 1800s (e.g., Davy and Berzelius) using electrolysis or reduction with potassium. Aluminum was obtained similarly from bauxite, but became available in large quantities in 1888 after development of the Hall process. While aluminum became a well-known item in commerce, boron was still very mysterious in the early 1900s.

Diborane

Alfred Stock (1876–1946)[229] began experimenting with boron, and starting around 1909 prepared some hydrides of boron, which he discovered to be volatile, flammable and toxic. In order to work with these compounds Stock invented techniques to manipulate

---

[229] Stock developed the modern system of nomenclature, e.g., iron(III) chloride (1919), and introduced the term "ligand" (1916).

and transfer volatile boron hydride in a vacuum system (e.g., the vacuum line) utilizing liquid-air cold traps and Teopler pumps. These systems typically utilize mercury as a working fluid and Stock developed mercury poisoning by the 1920s.[230]

The composition and physical properties of diborane are the result of its novel structure. In order to approach an octet of electrons in boron's 2s and 2p orbitals, the compound must either form an adduct or two-electron-three-center bridging bonds.

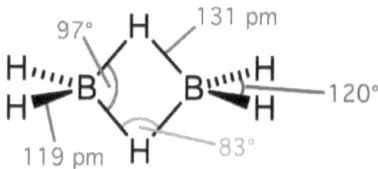

Drawing by Benjah-bmm27 and Esquilo,

---

[230] Mercury-powered vacuum lines (containing many pounds of mercury) were routinely used into the 1970s with some care concerning the disposition of mercury. However, the extreme problems in dealing with even minor mercury spills (i.e., the liquid penetrated into every crack and the vapors can be easily detected by cold-vapor atomic absorption) have led many institutions to abandon every form of mercury containing appliance. When I was a high school student, we made oxygen by decomposing HgO and the teacher let us take the droplets of mercury home to play with. No one died.

This type of bonding (common in organometallic compounds) was unknown in 1937 and based on electron diffraction and its molecular weight, the initial structure was assumed to be the same as ethane by Simon Bauer (1912-2013) a student of Pauling. However, the unusual reactivity of diborane was not consistent with the ethane structure. Hermann Schlesinger (1882-1960) wrote to Pauling (January 3, 1941)[231]:

> As a result of our work on the metallo borohydrides I definitely feel that a structure for diborane quite different from those generally proposed, would aid in correlating many of the observations we have made. ... Curiously enough I have just now received a reprint of a Russian article on hydrides of boron ... I gather from some of the formulae in the article that the author has come to a conclusion very similar to mine. The structure I have in mind is a bridge structure, in which the two boron atoms are joined to each other through an unusual type of hydrogen bond... ,

---

[231] Source Pierre Laszlo from Pauling Archive at Oregon State University, in Corvallis.

P a g e 319 | 483

Pauling was unmoved and responded (January 7, 1941):

> *I do not feel very friendly toward the structure which you mention in your letter for the diborane molecule. So long as the suggested structure remains vague and indefinite, it is not easy to say that it is eliminated by electron data or other data. However, the force constant for the B-B vibration is I think much stronger than would be expected for a structure of this type, in which there is no direct B-B bond.*

Pauling appears to have assumed that the B_{2u} and B_{3u} vibrations of the BH_2B structure were the vibration of a B-B bond.[232] Ironically, Schlesinger deferred to Pauling and continued with an incorrect model for years. In 1943, Christopher Longuet-Higgins published the correct structure[233] and created a firestorm of controversy and activity. The idea of the electron-

---

[232] See analysis by Parris for $Mg(CH_3)_2Mg$ system in G.E. Parris. *The solution composition of some organometallic reagents as inferred from spectroscopic and colligative property studies.* Thesis Georgia Tech pp. 98-110 (1974) available on the internet.

[233] H. C. Longuet-Higgins, R. P. Bell, *J. Chem. Soc.* (1943) 250-255.

deficient bridged bond was finally widely accepted by the NMR work of William N. Lipscomb (1919-2011).[234]

Sodium Borohydride

Hermann Schlesinger and his student Herbert Charles Brown[235] (1912–2004) were brought into the general support for the Manhattan Project in 1941 with the mission to find volatile compounds of uranium. Knowing that boron hydrides were very volatile and that aluminum hydride was relatively volatile, they set out to examine hydrides of uranium and these metals. That did not work out, but they discovered how to make sodium and potassium borohydride; and studied reactions of diborane, borane adducts and borohydrides with organic compounds. These results were not formally released until well after the war (*J. Am. Chem. Soc.*, 1953, 75 (1), pp 186–190). Work in the nuclear weapons program remained confidential into the Cold War.

---

[234] W.H. Eberhardt, B. Crawford, W.N. Lipscomb. The valence structure of the boron hydrides. *J. Chem. Phys.* 22: 989 (1954).

[235] Born Herbert Brovarnik.

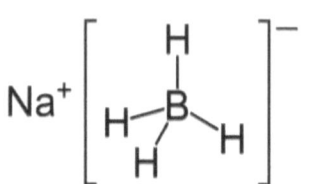

Drawing by Ben Mills, Pngbot
and Benjah-bmm27, source
Wikimedia Commons

Sodium borohydride (NBH) is prepared commercially by reaction of sodium hydride with trimethyl borate (250–270°C). It dissolves in ethers or protic solvents (water and alcohols) but reacts to produce hydrogen gas below pH 14 (e.g., with alcohols that have pKa less than 14). The reaction is slow in alcohols and it can be used for synthetic reaction in this solvent. It reduces aldehydes and ketones to alcohols but does not reduce carboxylic acids or their derivatives except acid halides (e.g., RCOX, X = Cl). However, hydrolysis products of $NaBH_4$ appear to polarize esters by forming adducts ($R_3B{\leftarrow}O{=}CORR$) which can be reduced with hydride from boron.

Lithium Aluminum Hydride

Interestingly, Schlesinger seems to have prepared lithium aluminum hydride (LAH) from the reaction of LiH with $AlCl_3$ after the war and it was reported in

1947. [236] It is produced commercially by reaction of sodium, aluminum and hydrogen at high pressure and then the sodium is exchanged with solid LiCl (precipitating NaCl from ether solvent). The final product contains about 1% LiCl.

LAH reacts violently with moisture in the air. It is normally used in diethyl ether solution although THF solutions are more stable. LAH is a more potent reducing agent than sodium borohydride and will reduce all acid derivatives.

13. Ferrocene

Like many other novel compounds, ferrocene was first prepared accidently in the late 1940s when dicyclopentadiene was passed over an iron catalysts (at Union Carbide). Samuel Miller and coworkers also prepared the material in an attempt to improve the ammonia catalyst. This material was recognized to be stable to water and acids, to melt at 163°C and to

---

[236] A.E. Finholt, A.C. Bond, H.I. Schlesinger. Lithium aluminum hydride, aluminum hydride and lithium gallium hydride, and some of their applications in organic and inorganic chemistry. *J. Am. Chem. Soc.* 69(5):1199–1203 (1947).

sublime without decomposition. In the meantime, Peter Paulson (1925-2013)[237] reacted cyclopentadienyl Grignard with ferric chloride hoping to couple and oxidize the organic moieties to form fulvalene.

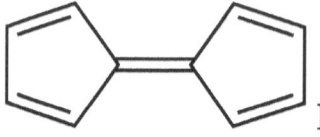 **Fulvalene**

Drawing by Eschenmoser, source Wikimedia Commons

Instead, they obtained a stable yellow powder, which they drew with sigma-bonds to carbon although they attributed the stability of the compound to the ability of cyclopentadienide to be aromatic (1951).

Ernest O. Fischer (1918-2007) in Germany and R.B. Woodward at Harvard became interested in this compound. Woodward (a senior professor) enlisted Geoffrey Wilkinson (1921-1996) (a junior professor) to work on the structural problem with him. In March 1952, Wilkinson and coworkers (including Woodward as senior author) submitted a short paper proposing the "sandwich" structure of the iron compound.[238]

---

[237] H. Werner. Peter Ludwig Pauson (1925–2013). *Angew. Chem. Int. Ed.* 53(13):3309 (2014).

[238] G.Wilkinson, M. Rosenblum,M. C.Whiting, R.B.Woodward. *J. Am. Chem. Soc.* 74:2125–2126 (1952).

Using principles of inorganic coordination chemistry, Fischer deduced (paper submitted June 1952)[239] that if the cyclopentdienyl groups each donated six electrons to Fe(II), the iron would achieve the stable electron configuration of krypton with essentially the same structure proposed by Wilkinson et al.

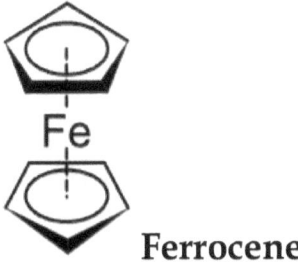

Ferrocene

Drawing by Roland Mattern, source Wikimedia Commons

It was subsequently determined by Woodward's coworkers that many typical aromatic substitution reactions could be applied to the rings without destroying the complex. Meanwhile Wilkinson and Fischer independently continued developments in metal-ocene chemistry and played a large role in founding the transition metal-organic complex

---

[239] E.O. Fischer, W. Pfab, Z. *Naturforsch.* B 7:377–379 (1952).

chemistry, which we will see developed in the next decades.[240]

Wilkinson and Fischer won the Nobel Prize for their work in 1973. Woodward was disappointed to the point that he wrote to the Nobel Prize Committee (October 26, 1973)[241]:

> "[The committee's decision to award the 1973 Nobel Prize without me]...*leaves me no choice but to let you know, most respectfully, that you have- inadvertently, I am sure- committed a grave injustice. ... Indeed, when I, as a gesture to a friend and junior colleague interested in organometallic chemistry, invited Professor Wilkinson to join me and my colleagues in the simple experiments which verified my structural proposal, his reaction to my views was close to derision....*"

Nonetheless, Wilkinson must have then read the Pauson paper (mentioned above) and he is cited thinking at the time "*...it can't be that, it must be a*

---

[240] H. Werner. At least 60 years of ferrocene: the discovery and rediscovery of the sandwich complexes. *Angew. Chem. Int. Ed.* 51:2–9 (2012).

[241] Quotes for Woodward and Wilkinson taken from *Organometallic Chemistry and Catalysis* By Didier Astruc p. 8. Springer-Verlag (2007).

sandwich." It is not clear whether Wilkinson was just recognizing that Woodward's proposal was correct or whether he had a flash of independent intuition. Regardless, he joined with Woodward, but Woodward points out that he (RBW) wrote the communication.[242]

14. Molecular Orbital Calculations by Organic Chemists

The 1933 series of papers by Pauling and co-workers and his book *The Nature of the Chemical Bond* (first edition, 1939) dominated thinking after WWII (1939-1945), while the efforts of almost every scientist in the world were turned to immediate practical problems. It is safe to say that most organic chemists were only generally familiar with this work through the early 1950s. Then they gradually became aware that molecular orbitals determine chemical reactions.

One of the problems from the chemists' standpoint was that molecular orbitals were indeed "molecular,"

---

[242] It is interesting to compare this episode to the situation with E.J. Corey that developed after Woodward's death.

characteristic of the molecule as a whole. Even the sigma bonds were treated by the physicists as parts of the molecule as a whole. In 1937, Coulson separated out the sigma localized bonding orbitals in his analysis of methane.

In 1947, Coulson managed to put the Hartree–Fock method into terms that organic chemists were likely to understand. He collaborated on a series of papers on the application of MO theory to conjugated systems.[243] They separated the molecules into the core (1s electrons), sigma-bonded electrons, and pi-bonding electrons. The first two categories were assumed to be independent and invariable. The pi-molecular orbitals were composed as linear combinations of the conjugated atomic p-orbitals. They note that there must be the same number of molecular orbitals (Ψj j = 1 to n) as atomic p-type orbitals (e.g., $2p_z$ or $3p_z$) contributing to the pi-system. They also considered hyperconjugation of alkyl substituents with the pi-system.

---

[243]C.A. Coulson, H.C. Longuet-Higgins. The Electronic Structure of Conjugated Systems. I. General Theory. *Proc. Royal Soc. A.* 191(1024): 39-60 (1947).

In analyzing the Schrodinger equations for these systems, they defined a term alpha (α_r) also known as the *Coulomb integral*, which is a measure of the stability of an electron associated with one atom (r) relative to the same term in benzene. And, they defined a *resonance integral* beta (β_{rs}) for and electron shared between two atoms (r and s). Referring to the work of Lennard-Jones (1937), they note that *"$2\beta_{rs}$ may be taken as the difference in energy between a pure single and pure double bond connecting r and s. It is a negative quantity whose magnitude increases along the series C=C, C=N, C=O."* They acknowledge that the overlap integrals are significant to absolute energy calculations, but by focusing on *relative energies* they are able to ignore these.[244]

The result is a series of linear equations in which alpha and beta are knowns and the coefficients (c_r) representing electron density on specific atom (r) is and unknown and the energy (E) of the orbital are unknowns. They solve the system of linear equations

---

[244] In his marvelous book *"Notes on Molecular Orbital Calculations"* (1961) John D. Roberts provides a graph from Mullikan (1952) that estimates the range of the overlap integrals (Sij) to vary from about 0.20 to 0.27 over the normal range of C-C bond lengths.

using the method of determinants (i.e., the secular determinant):

$$\text{Relative E (i.e. } \Delta E) = \begin{vmatrix} \alpha_1 - E_1 & \beta_{12} & \beta_{13}\ldots\ldots \\ \beta_{21} & \alpha_2 - E_2 & \beta_{23}\ldots \\ : & & \\ : & & \alpha_n - E_n \end{vmatrix} = 0$$

Once the values of E_j are obtained they can be plugged into the basic equation to calculate the corresponding coefficients (c_j) where

$$\Psi_j = \Sigma\ c_j(p_z)_j$$

And the absolute magnitudes of the coefficients are determined by normalization. Once the coefficients are known, the pi-bond order (P_{rs}) can be determined and the total electron density around each atom (q_r) can be calculated. The total energy of the system and the electrophilicity of each atom follow straightforwardly.

According to John D. Roberts (1918-2016), when he first set out to teach this material in 1950 to undergraduates, he realized that he did not understand it.[245] However,

---

[245] Roberts, J.D.: At the Right Place at the Right Time. ACS-Books, Amer. Chem. Soc., 1990. p.122.

with the help of a friend, he quickly learned and taught these concepts based primarily on the Coulson and Longuet-Higgins papers of 1947 (see above) throughout the 1950s at Cal Tech. Along the way, he actually became more rigorous than the work described above as data were published by other authors. By 1952, he was publishing his computational results in the *Journal of the American Chemical Society* and attracting the attention of organic chemists to particularly stable and unstable molecules and ions (e.g., cyclopropenyl cation). But, as we will see, the fraternity of organic chemists, was generally uninformed of the methods well into the 1960s.[246]

---

[246] I personally bought a copy of his 1961 book *Molecular Orbital Calculations* about 1969 while I was taking advanced organic chemistry as an undergraduate at NCSU. I still have it. I recently acquired a copy of Ian Flemings *Frontier Orbitals and Organic Chemical Reactions* (1976). Both are highly recommended.

15. The Origin of Life

The Urey-Miller Experiment

Harold Urey (1893–1981) was a student of G.N. Lewis who became interested in separating isotopes (1931) and discovered deuterium for which he received the Nobel Prize in chemistry in 1934. But, his interest extended into the abiotic synthesis of molecules that would ultimately assimilate into living organisms (i.e., abiogenesis). He speculated that the primordial atmosphere of earth would include ammonia, methane and hydrogen. In one of his lectures at the University of Chicago, his ideas caught the interest of a graduate student named Stanley Miller (1930–2007) who was working for Edward Teller, but who was actually more interested in organic compounds. Miller switched to Urey and created a system in which a Urey's atmosphere could be constantly recycled and exposed to electrical discharges. After a week or so, Miller was able to detect several amino acids by paper chromatography. The work was published under his name in *Science* in 1953. Miller continued this work and in various experiments using more modern analytical techniques. Thus, all the natural amino acids and more have been observed.

Protein Structure

Pauling, Robert Brainard Corey (1897–1971) and Herman Branson (1914–1995) published a series of papers using principles of structural chemistry to predict the three-dimensional shape of proteins. The most important paper was published in 1951, which led to the acceptance of the α-helix and β-sheet conformations of domains found in many proteins. Also in 1951, Frederick Sanger (1918–2013) and co-workers managed to determine the primary amino acid sequence in insulin. This was accomplished by successive degradation steps in which the N-terminal amino acid was labelled by nucleophilic aromatic substitution of the fluoride of fluorodinitrobenzene. The work of the Pauling group and the Sanger group put protein chemistry onto an organized basis. However, well into the 1970s, biochemistry textbooks were very vague about the composition of proteins and peptides and mechanisms of function were very mysterious. It was not until 1969 that Dorothy

Crowfoot Hodgkin[247] determined the three-dimensional structure of insulin by x-ray crystallography.

18. X-Ray Crystallography (1940s and 1950s)

Some Early Structures

Following the pioneering work of Bragg, various research groups began applying x-ray crystalography to solve the structure of complex organic compounds. In 1935 the structure of phthalocyanin and several of its metal complexes were determined by J. M. Robertson (*J. Chem. Soc.*, 615 (1935)). In 1948, the synthetic chemists were assisted by solution of the structure of strychnine. This application was even more important when Dorothy Crowfoot Hodgkin confirmed the beta lactam structure of penicillin (1949). In 1957, she published the structure of vitamin B12.

---

[247] Nobel Prize 1964.

George E. Parris   Copyright Claimed   Paperback 2019

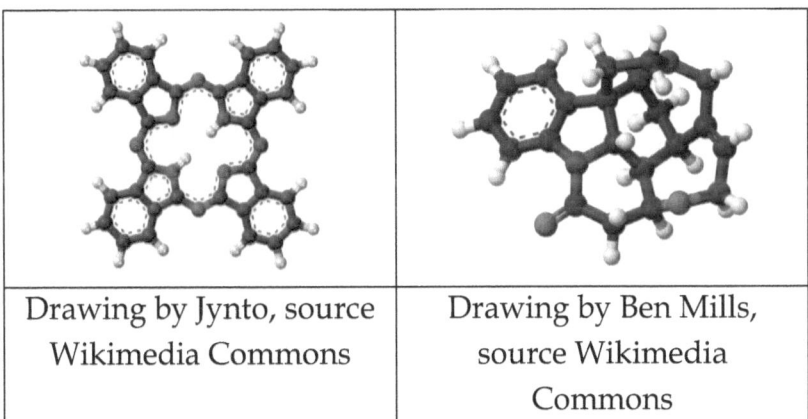

| Drawing by Jynto, source Wikimedia Commons | Drawing by Ben Mills, source Wikimedia Commons |

The Double Helix of Deoxyribose Nucleic Acid

The basic building blocks of DNA were known from the work of Phoebus Levene (1919). But, no one thought much about it because proteins seemed to be the most likely candidates for inheritable information. DNA was not recognized to be a polymer (capable of carrying information) until 1934 in the work of Torbjörn Caspersson (1910-1997). William Astbury (1898–1961) was the first to study the large-scale structure of DNA and recognized that it had a specific structural pattern (i.e., it is not just an amorphous solid) in 1937.

Work by biologists especially the group headed by Tomas Hunt Morgan (1866-1945), had associated chromosomes with inheritance of traits. His student Alfred Sturtevant (1891-1970) started identifying the locations of "genes" of the chromosomes in 1913 and Herman J. Muller (1890-1967) had shown that x-rays could cause mutations by damaging DNA in 1927 (Nobel Prize 1946). But chromosomes contained both DNA and proteins (nucleosomes). Thus, interest and research in DNA did not really become high priority until 1943 when Oswald Avery Jr. (1877–1955) and co-workers proved that DNA was the agent that carried genetic traits from one organism to another (essentially proving the Boveri–Sutton chromosome theory that biologists had been working with since 1905). This was the point where biology and chemistry merged and was the birth of "molecular biology."

After WWII (1945), three research groups (*none of them organic chemists*) became interested in the detailed structure of DNA and how it could carry inheritable and actionable information:

> Maurice Wilkins (1916–2004) and Rosalind Franklin (1920–1958) at King's College in London

> Francis Crick (1916–2004) and James Watson (1928-) at Cambridge[248]

---

[248] Crick was a physicist turned biologist and Watson was a phage biologist. Neither had any roots in organic chemistry.

Linus Pauling (1901–1994) at the California Institute of Technology

Pauling had become interested in proteins in the 1930s and with co-workers came up with the idea that the amino acids of many proteins were arranged in helical spirals (1951).

Crick and Watson became interested in Pauling's work and speculated that DNA might follow similar patterns and approached the problem applying structural organic chemistry concepts (i.e., building models). Watson had a background in x-ray diffraction, but they were not doing experiments.

Wilkins had settled at King's College after the war and one of his interest became DNA. He was able to isolate strands of DNA and began collecting x-ray data in 1950, but crystallography was not his specialty. In fact, Wilkins and Raymond Gosling had obtained x-ray data that was presented at a conference in Italy in the spring of 1951 which inspired Watson to pursue this topic. This success had also led to the hiring of Rosalind Franklin by the laboratory director without conferring with Wilkins. There seems to have been less collaboration between Wilkins and Franklin than between Wilkins and Watson.

Unlike Wilkins, Rosalind Franklin was an expert in x-ray diffraction[249] and she was recruited to King's College (1951) on the understanding that she would lead the study of DNA. But Wilkins thought she was hired as his assistant. The resulting clash of personalities was unfortunate for everyone except Watson and Crick. For the most part, Franklin worked alone on DNA, and grudgingly reported to Wilkins. This included providing raw data (x-ray patterns) to Wilkins.

On a visit to King's College, Watson (who understood crystallography at least as well as Wilkins) met with Wilkins. Not realizing the importance of the data provided by Franklin, Wilkins casually shared it with Watson. Watson immediately recognized the signature of a helical structure and carried this information back to Crick. Wilkins and Franklin plodded forward gathering data.

In early 1953, Pauling and Robert Brainard Corey (1897–1971) published a proposal for a DNA helix involving three intertwined strands of DNA with the

---

[249] She studies as a post-doc (1947-51) under Jacques Mering at the Laboratoire Central des Services Chimiques de l'Etat in Paris.

P a g e 338 | 483

bases on the outside while the phosphate "backbones" formed the central core.[250] Pauling shared his paper with Watson and Crick before publication.

Meanwhile Watson and Crick had realized from their models that certain base-pairs could fit together through hydrogen bonding in a double helix (a beautiful application of structural organic chemistry). This was then published in April 1953[251] with a brief acknowledgement of *"general nature of the unpublished experimental results of"* Franklin and Wilkins. This structure was soon shown to be correct and the unique feature was the base-pairing, which led to the critical biological implication:

> *"It has not escaped our notice that the specific pairing we have postulated immediately suggest a possible copying mechanism for the genetic material."*

Molecular biology **was born and organic chemists proceeded as though nothing had happened.**

---

[250] L. Pauling and R.B. Corey. A proposed structure for the nucleic acids. *Proc Natl Acad Sci U S A.* 39 (2): 84–97 (February 1953).

[251] J.D. Watson and F.H.C. Crick. A structure for deoxyribose nucleic acid. *Nature* 171, 737-738 (1953).

V. Modern Organic Chemistry (1955-1980)

In the post-war period, chemistry was seen as the key to advancement in consumer products (plastics, fibers, rubber, fuels, drugs, pesticides). With wealth and motivation in the US[252], there was rapid expansion of technology in academia and we see the saturation of chemical scientists. Specifically, because of the faith in science and the spur of the Cold War (including development of nuclear weapons and the brief lead held by the Soviet Union in space (earth-orbital missions, 1958-1963), almost any chemist with a PhD in 1955 could obtain an academic position if he (almost all males) wanted one.

In the U.S., R.B. Woodward became the superstar of organic chemistry. Much of the emphasis of "modern organic chemistry" (1955-1980 and beyond) as practiced by academicians that were trained in the period 1950-1970 continued to emulate his style and approach with much effort going into total synthesis.

---

[252] Not to mention the crippling of British, German, French, Russian and Japanese chemistry by WWII.

This period has saw great advancement in the tools available to chemists: Nuclear Magnetic Resonance spectroscopy moved from a novelty to an increasingly powerful tool with greater versatility (^{1}H, ^{7}Li, ^{13}C, ^{31}P, etc.), sensitivity (e.g., Fourier transform), and resolution (e.g., more powerful superconducting magnets). Gas chromatography similarly advanced in resolution (short packed columns replaced by long, open-tubular capillary columns), sensitivity and selectivity (advancement in detectors) and when coupled with mass spectroscopy and data management systems, GC-MS became an amazing forensic tool. The interest in large fragile molecules (by Woodward and others) also sparked parallel advances in High Performance Liquid Chromatograph. Mass spectrometers evolved in various ways aided by advanced electronics. Large resolving magnets were replaced by small quadrupole analyzes making GS-MS mobile. Inert atmosphere techniques with inert atmosphere boxes and specialized glassware became widely available. Radioactive and stable isotopes became routinely available and were used to resolve mechanistic ambiguities. In many cases, these advances allowed the scale of laboratory synthesis to be scaled down from grams to milligrams.

Interestingly, relatively few theoretical advances were made. Orbital symmetry and frontier orbitals became recognized as a driving force for most organic chemical reactions. Disputes broke out about the delocalization of charge in carbcations. Phenomena in which the time-scale for molecular dynamics (e.g., metathesis, exchange, and conformational changes) was shorter or longer than the time scale for observation (e.g., by NMR or IR) became useful for determining the rates and temperature dependence (hence, activation parameters and thermodynamics) of the processes.

Finally, the general disinterest of high-profile organic chemists and their academic departments in "molecular biology" (i.e., organic chemistry invented by biologists) is an important factor in the long term viability of organic chemistry. Academic chemistry departments generally retain their association with the physical sciences, rather than embrace the biological sciences. The academic organic chemists also eschew "material sciences" with separate departments (e.g., textile chemistry) frequently taking the lead in polymer research.

I was educated between 1965 and 1974 in chemistry and it is actually very remarkable that the material I teach to undergraduate students in 2019 is only incrementally different than the material I was taught 50 years ago.

1. Gas Chromatography

Archer Martin (1910–2002) worked in the same area as Richard Kuhn on vitamins and undoubtedly was very familiar with Kuhn's revival of Tsvet's methodology. Martin, thus, extended the technique of chromatography to different combinations of phases (liquid-liquid, liquid-paper, gas-solid, gas-liquid) in the 1940s. Gas-liquid chromatography was thus begun in 1950 and became an exotic (homebuilt) analytical instrument in college laboratories about 1957. Herbert C. Brown is said to have bought the first commercial GC (Perkin-Elmer model 154 GC system) for Purdue University in 1959.[253]

---

[253] Harold McNair. A History of Gas Chromatography: My Early Experiences. LCGC North America, Volume 28, Issue 2 (Feb 01, 2010).

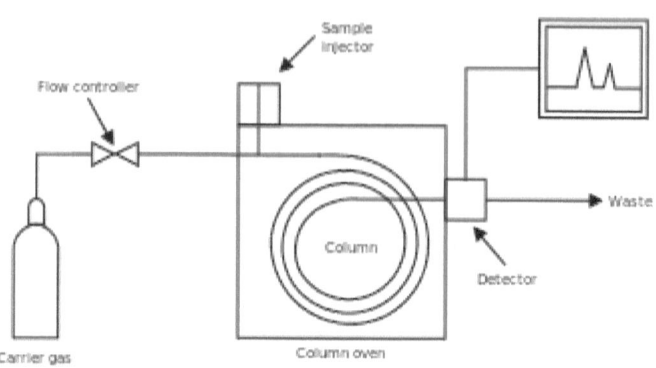

Drawing by Offnfopt, source Wikimedia Commons

Every element of the basic gas chromatography evolved as the techniques was made more sensitive, higher resolution, faster, etc. The two largest areas of improvement were in the columns, which were originally large bore (e.g., 1/8″ to 1/4″) steel or glass. Columns were commonly packed with irregular size and shape particles (e.g., brick dust, alumina) that had been given a thin coating of some high-boiling viscose oil. Temperature programming of the oven was initiated to accommodate samples containing both volatile and relatively low-volatility analytes. The detectors originally included thermal conductivity (TC, beginning 1950) that measured the heat capacity of the passing gas and then flame ionization detectors (FID, introduced about 1960), which gave orders of magnitude greater sensitivity. Petroleum companies led the development of these instruments and they were then introduced into academic laboratories. These general detectors then were supplemented with electron capture detectors (selective for halogens) and flame photometric detectors (selective for N, P and S). In the most modern versions, columns have been modified to open tubular capillary glass and the detectors are advanced mass spectrometers complete with data systems to capture mass spectra in cycles on the order of every 0.1 second during the chromatogram. The data systems allow storage and analysis of the data (e.g., comparison to standards for identification and quantification) before producing a graphic output (if indeed a graph is needed).

The move from packed columns to capillary columns (*circa* 1978) was based on issues with the theory of chromatography. As a packet of analyte-gas moves down the column, the molecules diffuse (both with and against the flow) causing peak broadening. Also the multiple random, circuitous pathways through a packed column caused broadening; and finally, the back pressure caused by the unavoidable restricted points along the packed column limited the length of the column (which ultimately limited the achievable resolution). Open-tubular capillary columns solved all these problems.[254] They also reduced sample sizes[255] and improved sensitivity. It should be noted that the efficiency of the packed column and the capillary column (theoretical plates, i.e., distillation events per unit length) are similar. But, the back-pressure problem limited typical packed columns to about 10-feet (2-3 m); whereas with the capillary columns, 100-m columns are feasible. The longer columns with rapid elution along a single channel resulted in an enormous improvement in resolution in the long capillary columns.

2. Nuclear Magnetic Resonance Spectroscopy

Once the Stern-Gerlach experiment was demonstrated, it became apparent that atomic magnetic moments would align themselves with an applied magnetic field and that if the temperature of the atom was sufficiently

---

[254] The small diameter glass tube is coated on the inside with a thin layer of liquid absorbent. This provides a pathway of uniform dimensions and properties.

[255] The vaporized sample may be "split" with only a small fraction (1%) actually taken into the column.

low, a preponderance of atoms would be in the lower energy spin state. That being the case, application of external electromagnetic energy (hv) of a frequency that coincided with the energy difference of the two magnetic states (ΔE) should result in net absorption of energy. This deduction was proven by Isidor Rabi (1898–1988) on ion beams in 1938. Work on radar during WWII (1939-1945) set the stage for the application of the technique to liquids and solids in the 1950s.

Exchange Phenomena and the Bloch Equation

Felix Bloch (1905–1983) was a contemporary student or colleague of many of the well-known European physicists who developed atomic theory in the 1920s. Being Jewish, he wisely abandoned Germany in 1934 and ended up at Stanford. After some time with the Manhattan Project, he moved to radar. In 1946, he published what are known as the Block Equations, which are differential equations describing the decay of net nuclear magnetization of a sample (i.e., relaxation of the system to the ground state) under influence of radio frequency stimulation. Relaxation can occur several ways (including chemical exchange of the nuclei of the magnetically skewed state with randomized nuclei) and the rate of relaxation can be related to the range of frequencies (i.e., signal width, typically measured as full-width at half-height) over which the energy exchange occurs. He and Edward Purcell (1912–1997) were primarily responsible for the theory behind NMR. This work was centered at Harvard and MIT; and it was originally seen as an adjunct to x-ray diffraction for studying solids. The NMR signals obtained from solids were very broad because the time scale of relaxation in the solid phase is very short.

Organic Chemist Discover NMR

To this point, the tool was in the hands of physicists, in particular, Purcell's graduate student, George Pake. Herbert S. Gutowsky (1919-2000) served four years in the US Army during WWII then began the PhD program. He had teamed with Pake to investigate several systems (including the structure of diborane). Gutowsky focused on wet chemistry tasks while Pake controlled the NMR instrumentation. Their studies of solids were inconclusive and they happened to examine liquids. To their surprise and pleasure, the signals from liquids were much narrower (because relaxation was much slower) and interpretable. This is where Gutowsky completed his PhD in 1948 and moved to the University of Illinois, where his principal duties were initially to provide infrared spectroscopy services to the organic research group headed by Roger Adams (1889–1971).

It took Gutowsky about two years to build an NMR spectrometer. Concurrently, Rex Edward Richards (1922-2019) at Oxford, who had earned his PhD in 1948 using infrared spectroscopy, was one of the first chemists to see the potential of NMR in Britain. He built an NMR spectrometer in 1949 (from military surplus radar parts) and published his first paper in 1951. Up until 1957, all NMR spectrometers were home-built.

Spin Coupling

The sharp signals obtained in the liquid phase by Gutowsky revealed that the signals for various protons in close proximity (typically three bonds) were coupled and created a local element of the magnetic field that was independent of the applied magnetic field. This coupling was manifest in a splitting of the signal into multiplets (e.g., doublets, triplets, quartets, etc.). The relative intensity of the components of the multiplet is determined by the statistical probability of the spin orientations of the nearby nuclear magnetic moments (i.e., nuclei) in the field (skewed by the mutual effects between the two sets of distinct protons). This work was first published in 1951.[256]

Chemical Shift

In 1950, several laboratories almost simultaneously realized that the frequency of NMR signals were affected by their chemical environments and this information spurred Gutowsky's interest in what is now known as the *chemical shift* of the signal relative to an internal standard (1953). The electron density around a nucleus tends to react to the applied magnetic field by producing a local field in the opposite direction that effectively "shields" the nucleus from the applied field. The stronger or weaker the shielding, the more the displacement of the observed signal from what would be anticipated for a "normal nucleus" (used as a

---

[256] H. S. Gutowsky, D. W. McCall, C. P. Slichter (1951). "Coupling among Nuclear Magnetic Dipoles in Molecules". *Physical Review* 84 (3): 589–90.

standard). Over the years, tetramethylsilane (TMS) has been chosen as a standard because: (i) with the slightly electropositive silicon pushing electron density into the methyl groups, its protons are more shielded that all compounds composed of C, H, O, N, S, P, and halogens such that a scale can be produced with most relevant chemical shifts on one side (down field) of TMS measured in units of displacement from TMS; (ii) TMS is chemically inert and soluble in most organic solvents; (iii) TMS is very volatile and can easily be removed from samples.

With the chemical shift and coupling constants understood, the power of NMR to solve structural problems for organic chemists was apparent. Thus, many laboratories wanted the technology. In 1957, Varian 40 MHz instruments became commercially available; and in 1961, the Varian A-60 (60 MHz) instrument was commercially available.

The technology has been progressively improved by providing more powerful magnets (up to 500 MHz instruments) and speeding up data collection and improving sensitivity through the use of the Fourier transform technique. In Fourier transform spectroscopy, instead of scanning one wavelength (or wave number) after another, the sample is "blasted" with a single pulse of energy and the signal that comes back is the result of a number of sinusoidal curves (each with a characteristic frequency, phase and amplitude). The Fourier transform de-convolutes these sine curves and translates the information into the parameters of chemical shift, line width and amplitude that chemists expect to see in NMR spectra.[257] This rapid data gathering allows multiple scans and summation of these scans amplifies the real signals while averaging out the random noise (improving the signal-to-noise ratio). Probes for various nuclei have become routine (especially carbon-13). The instruments have also incorporated electronics that automatically tune the instrument to each sample eliminating the tedious and time-consuming trial-and-error manipulating of knobs that graduated students in the period 1960-1975 dreaded.

Kurt Wüthrich (1938-)[258] pioneered use of the nuclear Overhauser effect (i.e., coupling of magnetic moments through space, not through bonds) to develop conformational models of complex biological molecules

---

[257] Richard Ernst won the 1991 Nobel Prize for this area.

[258] Nobel prize 2002.

in solution. This technique has become very useful in determining the conformation of large complex molecules (such as proteins).

Dynamic Systems

It turns out that application of the Block equations shows that for magnetic nuclei that are in different magnetic environments, but which can exchange their positions (e.g., the axial and equatorial positions of the cyclohexane ring), the exchange can average both the chemical shift and the coupling constants. For example, at ambient temperature (25°C), cyclohexane is usually observed as only a single sharp singlet because the ring is flipping faster than the NMR experiment is able to gather data, thus the NMR cannot resolve the actual interactions. However, if the sample is cooled to about -90°C, axial and equatorial protons are observable (100 MHz) as separate peaks (separated by a chemical shift differences of δaxial 1.12 and δequatorial 1.61). At this temperature, it is possible to resolve the germinal (Jae 13 Hz) and vicinal axial-axial (Jaa' 13.12 Hz) couplings; but it is not possible to resolve the smaller vicinal equatorial couplings (Jae' 3.35 Hz; Jee'

2.96 Hz).[259] The vicinal equatorial couplings can be resolved at -103°C (100 MHz).[260]

Note that the strength of these interactions (Jaa' >> Jee') are consistent with the efficient elimination (E2) of trans-periplanar constituents. In 1963, Martin Karplus[261] (1930-) published an equation for coupling of vicinal protons (on sp3 carbons) as a function of the dihedral angle (ϕ):

$$J_{(\phi)} = A \cos^2 \phi + B \cos \phi + C$$

Similar equations have been developed by Aksel Bothner-By (1921-).

---

[259] From spectrum of iodocyclohexane. F. R. Jensen, C. H. Bushweller, B.H. Beck. *J. Am. Chem. Soc.* 91(344): 3223 (1969).

[260] From spectrum of 1,1,2,2,3,3,4,4-octadeuteriocyclohexane at -103 °C. E.W. Garbisch Jr and M.G. Griffith, *J. Am. Chem. Soc.* 90:6543 (1968).

[261] M. Karplus, Martin. Vicinal Proton Coupling in Nuclear Magnetic Resonance. *J. Am. Chem. Soc.* 85(18):2870–2871 (1963).

3. Magic Acid and Organic Cations

Super Acids

The idea of super acids (i.e., an acidity greater than 100% sulfuric acid) had been introduced by James Conant in the early 1930s and it was understood that some salt-like compounds could be prepared by protonation of aldehydes and ketones by perchloric acid in acetic acid solution. This information contributed to the development of alkylation reaction of isobutene to produce isooctane in the 1940s.

While working for Dow Chemical in Canada, George Olah (1927-2017)[262] developed an interest in electrophilic reactions (e.g., Friedel-Crafts), which employ Lewis acids. By 1960 he was working at Case Western Reserve University and expanded the concept of a *super acid* by mixing a Lewis acid with Bronsted acids. The most common example is FSO_3H-SbF_5 sometimes called "magic acid."

---

[262] Born as Oláh György in Hungary; escaped to England after the 1956 anti-Communist revolt and moved to Canada where he went to work for Dow Chemical.

P a g e 355 | 483

Most atomic cations (e.g., Na^+) have a radius on the order of 10^{-10} m with most of the surface charge negative although the net charge is positive. Unlike all other atoms, when an electron is removed from hydrogen, all that is left is a positively charged proton with a diameter of about 10^{-15} m. Obviously the proton has enormous charge density and ability to polarize electron clouds around atoms and ions with inert gas electron configurations. Fluoride is about the least polarizable atom and we note that sulfur hexafluoride (SF_6) is a gas because the molecules have very little affinity for one another (very low polarizability). Thus, it is not surprising that SbF_6- would have the least possible affinity for a proton. Perchlorate similarly has a rather "hard" (non-polarizable) surface and is one of the strongest conventional acids along with sulfuric acid; but both of these acids have multiple bonds to oxygen that can be polarized. The same features of magic acid that make it a powerful acid also make it a poor nucleophile. The relative bulk of the conjugate base of the acid also ensures that its ion pairs will be loosely bound.

The power of magic acid was first discovered at an Olah group Christmas party in 1966. A paraffin wax candle was placed in the acid and it dissolved! They knew they were on to something. Experimentation revealed that even methane can be protonated: exchanging protons with the acid; producing H_2 and the very reactive methyl cation, which rapidly reacts with methane and other alkanes to yield more stable carbocations. Of course, any alkane, alcohol or carbonyl reacts with magic acid to produce carbocations that rearrange to the most stable form (typically tertiary).

Classical and Non-classical Carbon Cations

Up until this time (1970), organic chemists had used the term "carbonium ion" to refer to any sp2 hybridized carbon carrying a positive charge. But in the late 1960s, the possibility of stable bridged two-electron, three-center cations (initially termed "non-classical carbonium ions") arose. Olah's group published a series of papers that were interpreted in terms of the "non-classical" structure, which most organic chemists were quite willing to accept at face value.[263] However,

---

[263] In 1949, Saul Winstein (1912–1969) had first observed that solvolysis reactions of endo- and exo- 2-norbornyl -x compounds both gave racemic mixtures of exclusively the 2-exo-norbornyl-y products and had suggested a symmetrical 2-norbornyl cation.

H.C. Brown[264] (who was particularly concerned with what he called the "rococo period" of delocalized structures that were appearing in the chemical literature without experimental evidence) countered the Olah-group papers with similar experiments interpreted entirely as classical carbonium ions. There is no chemical reaction experiment that can definitively distinguish between rapidly rearranging classical carbocations and the non-classical structure. The only difference is what the free energy minima and maxima are in the intermediate (see figure below).

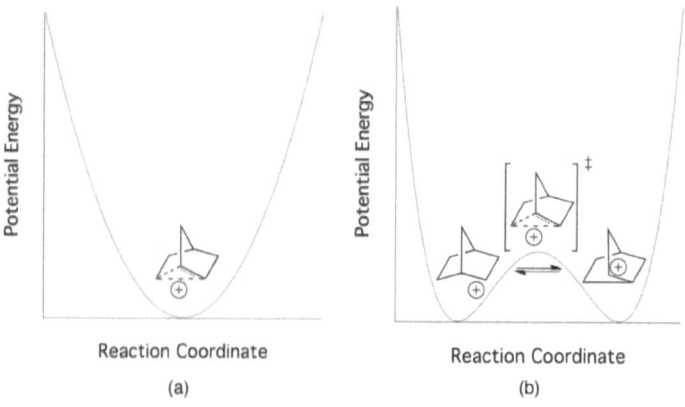

Drawing by Jcal730, source Wikimedia Commons.

---

[264] H.C. Brown and E.N. Peters. The selectivity principle and the question of the bridged structure of the 2-norborynyl cation. *Proc Natl Acad Sci U S A.* 71(1):132-4 (1974).

The time-scale of NMR spectroscopy is too slow to resolve the issue, however infrared spectroscopy and photoelectron spectroscopy as well as high-level (e.g., MP2/6-31G *ab initio*) calculations[265] support the non-classical (pentacoordinate carbon) species.[266] Olah's view eventually won out, and in order to cover this field more precisely, the classical ions were renamed as "carbocations" or "carbenium ions," with the term "carbonium ion" redefined to identify penta-coordinate positively charged carbon ions [CH_5^+].[267]

---

[265] R.A. Moss. The 2-norbornyl cation: a retrospective *J. Phys. Org. Chem.* 27(5):374–379 (2014).

[266] Although recent x-ray data support the non-classical structure in the solid phase (F. Scholz, F. et al. Crystal Structure Determination of the Nonclassical 2-Norbornyl Cation. *Science* 341(6141): 62–64 (2013)) I am not convinced that this solid phase structure can be assumed to represent what is present in solution. Lattice energy of the crystal may offset cation stabilities in solution.　IR is the most relevant tool for solution composition.

[267] G.A. Olah. Stable carbocations. CXVIII. General concept and structure of carbocations based on differentiation of trivalent (classical) carbenium ions from three-center bound penta- or tetra-coordinated (nonclassical) carbonium ions. Role of carbocations in electrophilic reactions. *J. Am. Chem. Soc.* 94(3):808–820 (1972).

4. Carbene Chemistry

Eduard Buchner began working with ethyl diazoacetate in 1885 and by 1903, he discovered that it would react with aromatic compounds to form cyclopropane derivatives.[268] He suspected that divalent carbon would be an intermediate. Staudinger also made cyclopropane derivatives by reacting diazomethane with olefins (1912). In the 1940s, physicist Gerhard Herzberg (1904–1999)[269] was studying spectra of molecules in outer space. Initially he mis-identified the spectra of C_3 as CH_2, but when this error was resolved, Herzberg generated CH_2 by flash photolysis of diazomethane (1949). However, the fact that the ground state is a triplet (with H-C-H bond angle 136° and C-H bond length 1.078 A) was not established until 1971.[270] There are also two very short-

---

[268] E. Buchner and L. Feldmann. *Diazoessigester und Toluol.* *Berichte der deutschen chemischen Gesellschaft.* 36(3):3509–3517 (1903).

[269] Fled Germany to Canada in 1935 with his Jewish wife.
[270] G. Herzberg and J.W.C. Johns. *J. Chem. Phys.* 54:2276 (1971). See discussion in the Nobel lecture of Herzberg (1971).

lived singlet states produced by photolysis of diazomethane (bond angles 140 and 104 degrees). Dichlorocarbene is easily generated by the action of strong base on chloroform.[271] It appears to be a singlet (Cl-C-Cl bond angle 109°). William von Doering (1917–2011) showed that it reacted with olefins in 1954[272] and used dibromocarbene to facilitate the conversion of an olefin into an allene by chain extension in 1958.[273]

Simmons-Smith Reagent (1958)

Howard Simmons, Jr. (1929–1997) and Ronald D. Smith are two of the most obscure modern organic chemists. Almost nothing is written about them or by them. Working in industry (DuPont Central Research) instead of academia, quiet professionalism seems to be the

---

[271] J. Hine. Carbon dichloride as an Intermediate in the basic hydrolysis of chloroform. a mechanism for substitution reactions at a saturated carbon atom. *J. Am. Chem. Soc.* 72(6):2438-2445 (1950).

[272] W. von E. Doering and A.K. Hoffmann. The addition of dichlorocarbene to olefins. *J. Am. Chem. Soc.* 76(23):6162–6165 (1954).

[273] W. von E. Doering and P.M. Laflamme. A two-step of synthesis of allenes from olefins. *Tetrahedron* 2(1-2):75-79 (1958).

preferred traits of chemists. The reaction that carries their name is the generation of a reagent assumed to be "IZnCH₂I" (formed from the reaction of methylene iodide with zinc-copper couple), which undergoes reactions that appear to involve carbene (:CH$_2$), specifically, the formation of cyclopropanes with olefins. Normally, the reagent attacks double bonds from the less hindered side, but the presence of a nearby –OH group appears to coordinate with the zinc and facilitate formation of the cyclopropane on the same side as the –OH. Reactions with unstable cis-olefines that would be expected to yield trans-substituted cyclopropanes in a two-step addition suggest that the cyclopropane ring is formed essentially in a concerted fashion. There appears to be much more work to be done to sort out the solution composition and reactivity of the reactive zinc species.

CH_2I_2, Zn(Cu), Et$_2$O → Bicyclo[4.1.0]heptane + ZnI$_2$

The Simmons-Smith Reaction

Drawn by V8rik, source Wikimedia Commons

5. Mass Spectroscopy

The first application of mass separation technology by applying magnetic fields to streams of moving atomic ions in the gas phase, was in 1913 by J.J. Thomson. It immediately yielded the important result of revealing isotopes of neon. Francis William Aston (1877–1945) continued this work at the Cavendish Laboratory after WWI and in 1919 build the first practical mass spectrometer, which he steadily improved over the next three years while identifying over two-hundred isotopes. In his book on the subject (1913), Thomson wrote[274]:

> *"I have described at some length the application of positive rays to chemical analysis; one of the main reasons for writing this book was the hope that it might induce others, and especially chemists, to try this method of analysis. I feel sure that there are many problems in chemistry which could be solved with far greater ease by this than by any other method. The method is surprisingly sensitive – more so even than*

---

[274] S.H. Bauer. Mass spectrometry in the mid-1930's: were chemists intrigued? *Journal of the American Society for Mass Spectrometry.* 12(9): 975–988 (2001).

that of spectrum analysis, requires an infinitesimal amount of material, and does not require this to be specially purified: the technique is not difficult if appliances for producing high vacua are available"

Meanwhile, Arthur Jeffrey Dempster (1886-1950) recognized the technical issues associated with obtaining accurate mass measurements including variation of the velocity of the ions, convergence of the selected ions to a small spot and detection of the ions over a wide range of intensities. He then developed a more practical arrangement for observing molecular fragmentation patterns with much higher accuracy in measuring m/e, this allowed him to resolve the two principle isotopes of uranium (U-235 and U-238) in 1935. Obviously at this time, the focus was on radioactivity and in 1937 three isotopes of potassium were isolated so that their relative radioactivity could be measured. Indeed, when the possibility of building atomic bombs from enriched isotopes became apparent, secrecy fell over these developments that followed through WWII and into the cold war (1950s). The Y-12 plant at Oak Ridge, TN used electromagnetics to separate uranium isotopes (1944-). In Germany, Manfred von Ardenne (1907–1997) developed similar technology (scanning electron microscope) and applied it to isotope separation in Germany (1939-45) and after

WWII in Russia. Thus, academic use of mass spectroscopy for organic chemical analysis was delayed until the governments decided that such research would not compromise national security.

The other reason that organic chemists did not pursue mass spectroscopy was that they were not inclined to develop the theoretical understanding of fragmentations of organic molecules. In the 1930s and 1940s, organic molecules strongly energized by the ionization process, seemed to fragment randomly. The first hint that regular mechanisms of fragmentation might be possible came from Australia in 1954 where A. J. C. Nicholson observed that a gamma hydrogen (H.) shift accompanied beta cleavage of certain compounds. However, this was not immediately recognized as a common mechanism of fragmentation, and the utility of the MS technique for structural determination of organic molecules was still not obvious.

Fred McLafferty (1923-), a decorated WWII infantryman, earned his PhD in 1950 and became interested in using the mass spectral technique as a detector for gas chromatography. The total current of ions generated by mass fragmentation is very similar to the effects obtained in a flame ionization detector.

Thus, in the mid-1950s at Dow Chemical Company, he made the first GC-MS interface. While the MS was intended as a detector for GC, the GC became a convenient way to introduce samples into the MS. McLafferty was, thus, able to quickly examine the mass fragmentation patterns of numerous organic molecules and established that the rearrangement observed by Nicholson was quite general. When he published his results in 1959[275], the power of mass fragmentation analysis as a reproducible, mechanistically controlled phenomenon became apparent to a wide audience of organic chemists. The instruments, however, were expensive, cumbersome and required substantial maintenance, which slowed their penetration into academic laboratories. It was not until the 1970s that GC-MS became routinely available.

---

[275] F. W. McLafferty. Mass Spectrometric Analysis. Molecular Rearrangements. *Anal. Chem.* 31(1): 82–87 (1959).

6. Crown Ethers, Host-Guest Compounds and Phase-transfer Catalysis

Charles Pedersen (1904–1989)[276] worked most of his career for DuPont. In 1967, he got the attention of academic chemists by publishing a method of synthesizing cyclic-polyethers (i.e., "crown ethers")[277] and showing that they complexes alkali metals selectively and stably.[278] When this occurs, the surface charge density of the complex ion is so low that solubility in relatively non-polar solvents is feasible (i.e., the lattice energy of the complex cations with anions is less than the dispersion forces between the complex ion and the solvent.)

The catalytic effects of crown ethers, thus, are much like those of tetraalkylammonium or phosphonium

---

[276] Born in Korea from a Japanese mother and Norwegian father, came to the US in 1922, received MS from MIT in 1927.

[277] "the cyclic tetramer of ethylene oxide" had been reported in a British patent in 1957 and was cited by Pedersen who was very familiar with the patent literature.

[278] C.J. Pedersen. Cyclic polyethers and their complexes with metal salts. *J. Amer. Chem. Soc.* 89(26):7017–7036 (1967).

salts, which allow anionic nucleophiles (e.g., cyanide) to pass into non-polar and non-protic media where they react rapidly in substitution reactions (Sn2).

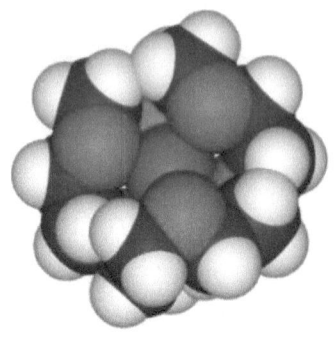

18-crown-6-potassium[279]

Drawing by Benjah-bmm27, source Wikimedia Commons

Donald James Cram (1919–2001)[280] had become well known in the early 1950s for his work on asymmetric induction by existing adjacent asymmetric centers in

---

[279] Crown ethers do not have functional groups normally associated with toxicity; however, their ability to selectively tie up potassium cation makes them toxic (and possibly mutagenic) because potassium ion is critical to many cellular processes and is associated with the phosphate ions of DNA and RNA. 18-crown-6 will even facilitate the liberation of electrons from potassium metal (similar to liquid ammonia) causing reactions of free electrons (e.g., release of hydrogen by reduction of hydrocarbons).

[280] PhD from Harvard in 1947 under Louis Fieser (1899 –1977).

molecules that influence the transition state at a reacting center (e.g., an asymmetric center adjacent to a carbonyl as it is converted to an alcohol). In the 1970s his interests turned to host-guest molecules and he was an author of the 1977 *Organic Synthesis* paper for 18-crown-6 (*Org. Synth.* 1977, 57, 30).[281] His principal interest was with a closely related group of compounds that form host-guest complexes including sphereands defined as *"systems of ligands organized prior to complexation so that the orbitals of unshared electron pairs of the binding sites line a roughly spherical cavity enforced by a support structure of covalent bonds."*[282] Cram had become aware that enzymes preform complex transformation with complete asymmetric control because they were pre-formed in a structure that not only brought all the functional groups together, but also expedited the reaction by nullifying the entropy penalty normally

---

[281] G. W. Gokel, D. J. Cram, C. L. Liotta, H. P. Harris, and F. L. Cook, *J. Org. Chem.*, 39, 2445 (1974).

[282] D.J. Cram et al. Host-guest complexation. 35. Spherands, the first completely preorganized ligand systems. *J. Am. Chem. Soc.* 107(12):3645–3657 (1985).

encountered in bringing reactants together in solution.[283]

7. High Performance Liquid Chromatography

In the 1940s and 1950s synthetic chemists moved towards total synthesis of very complex natural products such as derivatives of penicillin and vitamin B12. These compounds were not feasibly for gas chromatography and the early work was done laboriously through slow column chromatography. I recall when I arrived at Georgia Tech in the summer of 1969, I came across a laboratory in the old chemistry building near Grant Field where several glass columns over 10-feet tall and about 3-inches in diameter were slowly eluting something using gallons of solvent. In many laboratories, low pressure of air applied to the top of a chromatography column was used to enhance the gravity flow of the eluent.

In 1970, Csaba Horváth (1930–2004) coined the term HPLC for high pressure (500 psi) liquid chromatography. The improvement in speed and

---

[283] D.J. Cram. Preorganization — From Solvents to Spherands. *Angewandte Chemie* International Edition in English. 25(12):1039–1057 (1986).

resolution (from minimization of random diffusion) was remarkable and soon the Waters Company had made many changes to the technique to accommodate pressures up to 6000 psi. By the late 1970s, the instrumentation was entering academic laboratories. (I saw the first system at the Food and Drug Administration in 1977.) The particle size of the stationary phase (i.e., surface area/unit mass increases) and the pressures are still going up with experiments in the range of 100,000 psi in the 2000s.

There was a story circulating in the mid-1970s that R.B. Woodward first saw the instrumentation and immediately acquired a system. The story goes that some of Woodward's students completed separation of intermediates in total synthesis (which had taken years) to do be redone in days.[284]

---

[284] I think I heard this from a Waters salesman making a pitch at the FDA.

8. The Pharaoh[285] and Stereochemical Control

The legend

There have been and always will be many people with large egos in academia. Woodward had a large ego, but he had mastered the ability to control his ego within a range where most other chemist admired him rather than feared or envied him. He did this by enormous personal effort supported by single-minded interest. It is clear that from a very early age, "he ate and slept" chemistry: It was his job; it was his hobby; it was his muse. His seemingly spontaneous lectures were undoubtedly rehearsed. In some cases, Woodward asked questions about topics that he had already solved, just to amaze the audience with his "spontaneous" solution.

He was noted for drinking a pitcher of daiquiris and chain-smoking through formal presentations of his work. He is quoted as having said, "I never met a competent organic chemist who didn't smoke." His presentations were generally made on a blackboard with multi-colored chalk. His presentations were

---

[285] R.B. Woodward was once carried to a speaker's platform in a sedan chair carried by colleagues and students.

incredibly long. He demanded complete devotion from his students and post-docs to their work. He is said to have had a passion for the color blue (his tie, his parking place, the ceiling of his office). I would bet money that he simply wore the blue tie a few times and students decided it was one of his trademarks and provided the other blue effects, which he happily went along with as part of his legend. It is hard to imagine Woodward on a campus now: His trademarks would be prohibited; his lectures would be slow; his habits would be considered boorish; he would be considered overbearing; and his single-mindedness would be considered a psychosis. And, of course, he had all the social biases so frowned upon in academia today.[286]

The Woodward-Hoffmann Rules

One of the last/most-recent major additions to the understanding of chemical reactions was provided by the Woodward-Hoffmann rules. These rules explain the observed stereochemistry of concerted rearrangements and some bimolecular reactions under thermal and photochemical conditions. The initial publication was in 1965, but like many good ideas, they

---

[286] You can actually watch videos of some of his lectures on www.youtube.com. For example:
https://www.youtube.com/watch?v=YvEB05xdAy4

were built on earlier concepts and convoluted with some claims of priority.[287]

Harvard, of course, is staffed by talented and dedicated scientists. Woodward, as discussed above, was famous for his lengthy seminars and his technique of posing questions to his audience and discussing the possibilities (even when he knew the answer). No doubt, he got a number of good ideas from these discussions; but since he (i.e., his group) was the only one pursuing a particular synthesis at Harvard, no one was phased and he was sharing his experience with them. It is clear that Woodward left few things to chance.

However, Woodward was not in the theoretical chemistry business and according to Elias James "E.J." Corey (1928-) Corey[288] the root of the (Nobel Prize-winning) rules actually came from him in a private discussion. Woodward had been fascinated by the Diels-Alder reaction and related reactions in which pi-bonds and sigma-bonds were formed from one another all of his career. He no doubt recognized that under

---

[287] R.B. Woodward, R. Hoffmann. Stereochemistry of Electrocyclic Reactions. *J. Am. Chem. Soc.* 87 (2): 395 (1965).

[288] A student of John C. Sheehan at MIT. Who joined Harvard in 1959.

various conditions clear stereo chemical preferences were observed when no preference was expected.

In May 1964, Corey (age 35) was a relatively new colleague (Woodward was 47 and had been at Harvard for 26 years and was in line to receive the Nobel Prize in 1965). According to Corey, in a private conversation on the evening of May 4th 1964, Woodward asked him how he would explain certain stereochemical observations. Corey was probably closer to modern molecular orbital theory than Woodward was; and after some discussion, he suggested to Woodward that orbital symmetry was probably the driving factor. Woodward seemed pleased and Corey apparently went away dreaming that he and Woodward would collaborate on the development of these ideas:[289]

> "On May 4, 1964, I suggested to my colleague R. B. Woodward a simple explanation involving the symmetry of the perturbed (HOMO)[290] molecular orbitals for the stereoselective cyclobutene / 1,3-butadiene and 1,3,5-hexatriene/cyclohexadiene conversions that provided the basis for the further

---

[289] Transcript from Priestley Medal acceptance speech published in *Chem. Eng. News* 82(13), 42–44 (2004).

[290] "HOMO" is Highest Occupied Molecular Orbitals.

development of these ideas into what became known as the Woodward-Hoffmann rules."[291]

But, Woodward must have already been thinking about the molecular orbitals as a source of the selectivity because he had apparently already engaged a post-doc Roald Hoffmann (1937-)[292] (one of Lipscomb's students

---

[291] Some sources say that Woodward was skeptical about the idea at the time. However, *if Woodward were seriously skeptical,* why would Corey anticipate collaboration? Moreover, if Corey were certain of his idea and Woodward had rejected it, he should have immediately (that night) written a note to the editor of some journal describing the idea and mentioning the conversation with Woodward. This note could be shared with Woodward before submission to a journal for comment. He would have either received Woodward's blessing or forced Woodward to consider collaboration. This is not too bold a move for Cory to make. I personally as a graduate student presented the idea that solvation of the principal species of the Grignard equilibrium was the deciding factor in determining the position of the equilibrium to one of E.C. Ashby's seminars and received skepticism and negative comments from Ashby (my thesis advisor). I wrote such a letter to the editor of *Inorganic Chemistry* and sent it in. When it was accepted for publication, I figured I should make it known to Ashby, he was (rightly) disturbed and I voluntarily withdrew the letter. I was somewhat disturbed to latter see similar verbiage appear in Ashby's publications with other students. But I figured I had stepped over the line and was more interested in completing my degree than fighting my boss.

[292] Born Roald Safran in Poland.

at Harvard) who was about to receive his PhD after working on the Extended Hückel method[293] of analysis of pi and sigma orbitals. We know this because Hoffmann mentions making a notebook entry about the idea (from a discussion with Woodward and D.E. Applequist) on May 5th 1964.[294] (Hoffman says this date may have been added latter and acknowledges that Applequist does not recall any earlier discussions with Woodward.) In any event, Hoffmann acknowledges that Corey seemed to have assumed that he was in some way involved with the work of Woodward and Hoffman in his discussions with Hoffmann during the summer of 1964; to the point that Hoffmann felt compelled to ask Woodward about the authorship of their first paper. But, when asked if Corey was involved, Woodward merely said "no."

The impression I get from all of this is that Corey was not brave enough to approach Woodward, but instead tried to get the message of his anxiety to Woodward via Hoffmann. It seems that Corey did at the time believe

---

[293] A semi-empirical quantum chemistry method of determining orbital energies.

[294] R. Hoffmann. Woodward–Hoffmann Rules: A claim on the development of the frontier orbital explanation of electrocyclic reactions. *Angew. Chem. Int. Ed.* 43:2–6 (2004).

he had made relevant disclosures to Woodward; but at the time, the combination of the unknown importance of the Woodward-Hoffman rules and the fear of challenging Woodward directly prevented Cory from bring the issue up...until after Woodward died (1979). Indeed, Corey did not raise the issue formally with Hoffman until 1981 (when Hoffmann was set to receive the Nobel Prize) and he did not raise the issue publicly until 2004 (when he (Corey) had likely "peaked" in his academic success). Overall, I think that Hoffmann (*Angew. Chem. Int. Ed.* 43:2–6 (2004)) expresses the frustration caused by Corey's inactions and actions well. While Woodward may have forestalled Corey (which appears to have been easily done), there is no evidence that Corey independently discovered the idea and he certainly did not develop it. In fact, I am about to show that there was another source (duly cited by Woodward) that overshadows Corey's claim of contribution.

What the Theoretical Chemists Knew

When I examine the literature leading to May 1964, I can only believe that E.J. Corey's claim is an admission of broad ignorance. Theoretical chemists and spectroscopists, had been working with correlation diagrams for years. In 1953, Arthur Walsh (1916-1977) published papers on the energy of molecular orbitals as a function of bond angle in which he suggested that they could predict the stability of small molecules. In addition to energy, these diagrams link specific orbitals of the reactants to orbitals in the products based on symmetry.

In the 1950s the Hückel molecular orbital method (HMO) had been developed by theoretical chemists and popularized by John D. Roberts, so that most organic chemists were aware that conjugated systems could be represented by linear combinations of atomic p-orbitals, and that the energies of MOs and electron densities for each atom in such a system could be calculated.

Luitzen Johannes Oosterhoff (1907-1974)[295] of Leiden University was a prolific theoretical chemist who was frequently on the cutting edge of molecular orbital

---

[295] E. Havinga and J.L.M.A. Schlatmann. Remarks on the specificities of the photochemical and thermal transformations in the vitamin D field. *Tetrahedron*. 16:146-152 (1961). Specifically, p.151.

Abstract: "A discussion is given of some mechanistic aspects of the photochemical and thermal isomerization reactions occurring in the vitamin D field. Possible explanations of the stereochemical specificities of these reactions are tentatively put forward."

phenomena in chemistry. Organic chemists in his institution brought problems to him for comment and clarification. E. Havinga and J.L.M.A. Schlatmann had been investigating the synthesis of Vitamin D. They had noted that photochemical and thermal reactions yielded different products and asked Oosterhoff for suggestions as to what was going on. They summarized his comments in their 1961 paper as follows:

> *As Prof. Oosterhoff pointed out, another factor that possibly contributes to the stereochemical difference between the thermal and the photo induced ring closure may be found in the symmetry characteristics of the highest occupied π orbital of the conjugated hexatriene system. In the photo excited state this highest occupied orbital is antisymmetric with regard to the plane that is perpendicular to the bond 6,7 making 'syn' approach less favourable.*[296]

---

[296] The massive energies of Woodward were focused on the synthetic literature (indeed, when I was in school every organic grad student was typically expected to scan all the current articles in about 10 high-profile journals (and answer questions on cumulative exams) and it is likely that he read this paper in 1961. Corey may have unwittingly reminded Woodward of this detail. Regardless, Cory's revelation cannot have been more insightful than this article and once this paper was available to Woodward, the conversation with Corey in 1964 was not relevant.

It is not clear how Oosterhoff came to this conclusion, but he was clearly aware of orbital symmetry and believed it was a factor in controlling chemical reactions. However, he never published a methodology for its application.

In 1963, Hoffmann was developing what is known as the extended Hückel molecular orbital method, which includes both the pi and sigma bonding orbitals are considered. He developed a localized matrix element for the sigma bonds (e.g., for the bond between atoms r and s, H_{rs}) which was a product of the average ((H_{rr} + H_{ss})/2) alpha terms for atoms r and s, and the overlap integral (S_{rs}) and the empirical Wolfsberg-Helmholtz constant (i.e., 1.75). This mathematics was obviously appropriate of situations where pi bonds were being converted to sigma bonds.

What the Organic Chemists Observed

In 1958-9, two reports[297] of the thermal ring-opening of 3,4-di-substituted cyclobutenes indicated that the reaction was stereo-specific. We have already noted the observations on stereo chemical changes in the

---

[297] E. Vogel. *Liebigs Ann. Chem.* 1958 615:14 and R. Crigee and K. Noll. *Liebigs Ann. Chem.* 1959, 627:1.

synthesis of vitamin D (see above) in 1961. And, Woodward ran into a similar situation in the synthesis of vitamin B_{12}, which peaked his interest in the phenomenon. Corey's comments on May 4th 1964, may have triggered Woodward's research, but his contribution was superseded by Oosterhoff's. Woodward might have asked Corey to collaborate, but Hoffmann was much better versed in the calculations that needed to be done and Corey (with his own research program) was not a logical choice to collaborate with. Hoffmann was needed to crunch the numbers.

The Resulting Rules

Stated tersely, the following rules can be applied by even the most mathematically challenged organic chemist (for electrocyclic reactions):

(1a) Conjugated systems with 4n + 2 electrons have the same polarity of the terminal p-orbitals of the highest occupied molecular orbital (HOMO) (↑ ↑) and can thermally (ground state) form a sigma bond (→←) between the terminal carbons only by disrotatory twisting of the first and last double bonds.

(1b) For the same (4n +2) systems, promotion of an electron to the first excited state (i.e., HOMO) reverses the polarities of the terminal p-orbital lobes (↑ ↓) and requires conrotary twisting during sigma bond formation (→←), which leads to reversal of stereospecificity.

(2a) For similar systems involving only 4n electrons, the signs of the lobes of the terminal p-orbitals are opposite in the HOMO (↑ ↓) in the ground state and can achieve positive overlap in the transition state by thermal reaction state by con-rotary twisting (→←).

(2b) For the same (4n) systems, promotion of an electron to the first excited state (HOMO) reverses the polarities of the terminal p-orbital lobes (↑ ↑) and requires disrotary twisting during sigma bond formation (→←), which leads to reversal of stereospecificity.

Similar reasoning applies to other types of concerted pericyclic processes: cycloadditions (e.g., Diels-Alder), sigmatropic reactions, group transfer and chelotropic reactions.[298]

---

[298] There are many excellent examples provided graphically in Wikimedia Commons. Although it is open access, I will refer you there, rather than insert large amounts of this material here.

Retrosynthetic Analysis

E.J. Corey, regardless of his involvement with orbital symmetry, is well known and successful in his own right.[299] There are several reagents that are important to organic synthesis that came from his work. For example, the Corey-Suggs reagent (pyridinium chlorochromate, PCC) is conveniently used to oxidize alcohols to ketones and aldehydes. He also developed selective reagents for formation of protecting groups for sensitive functional groups in complex syntheses. But, I believe that Cory's most lasting contribution is the art of retrosynthetic analysis. With today's powerful spectroscopic and chromatographic techniques for determining structures, total synthesis for the purpose of proving structures seem to be less important than in 1950. But what Corey has championed is the idea that the process of synthesis of complex molecules (of known structure) can be rendered systematic in a step-by-step deconstruction of the "target" into an assemblage of "synthons" known

---

[299] Most people regard his claims of association with orbital symmetry, as an episode that diminished his reputation much more than credit for his claimed contribution ever could have enhanced it.

as a "retrosynthetic tree." The approach lends itself to artificial intelligence (computerized) analysis of synthetic problems.

Cahn–Ingold–Prelog Priority Rules

Emil Fischer (1891) succeeded in resolving many of the "hexose" compounds and their optical rotations (d/l or +/-) could be measured, but he was unsure of their absolute configuration. This problem became more serious as work was done on other types of compounds (i.e., not related to hexose). Martin André Rosanoff (1874–1951)[300] was the first to attempt this classification (which is represented by the modern D/L system)[301], but this approach was still very limited. The absolute configurations were not confirmed until 1951.

With the advent of total synthesis of complex biological molecules in the 1940s, it was realized that a simple, unambiguous system needed to be introduced to assign absolute configurations to each asymmetric center.

---

[300] Rosanoff worked for Edison circa 1903 and is the source of the Edison quote: *"Hell! there ain't no rules around here! We are tryin' to accomplish somep'n!"*

[301] M.A. Rosanoff. On Fischer's classification of stereoisomers. *J. Am. Chem. Soc.* 28:114 (1906).

Vladimir Prelog (1906–1998), who became interested in synthesis of biological compounds, was probably the first to see the need and Robert Sidney Cahn (1899–1981), and Christopher Ingold (1893–1970) were working on problems that required appreciation of absolute configuration. Out of their collaboration, a comprehensive system (R/S) of nomenclature of absolute configuration evolved (1966).[302] This system of prioritization (based on atomic numbers of substituents) is also used in the E/Z nomenclature on symmetry around double bonds.

9. Transition Metal Catalysis

Olefin Metastasis

Olefin metastasis is not a reaction that was anticipated based on classical structural ideas of chemistry. It seems to involve an improbable structural rearrangement;

---

[302] R.S. Cahn; C.K. Ingold; V. Prelog. Specification of Molecular Chirality. *Ang. Chem. Int. Ed.* 5(4):385–415 (1966).

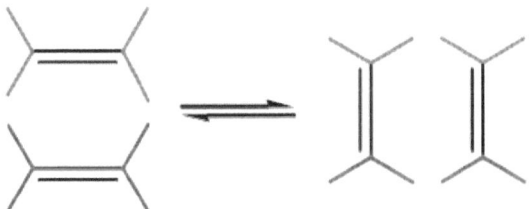

Drawing by Wjousts, source Wikimedia Commons

And after introduction of the Woodward-Hoffmann rules (late 1960s) the probability of a 2+2 cycloaddition intermediate leading to olefin metastasis seemed even more remote. Nonetheless, Ziegler had observed unexpected effects of transition metals on polymerizations; and during the period 1950-1970, industrial chemistry patents periodically appeared with unexpectedly novel rearrangements of olefins.[303] In 1971, Yves Chauvin (1930–2015) proposed what turned out to be the correct mechanism for these reactions.[304] The catalysts for these reactions is formed by displacement of a ligand (e.g., phosphine, PR_3) from a second or third row transition metal (e.g., Mo, W, Ru,

---

[303] Wikipedia (https://en.wikipedia.org/wiki/Olefin_metathesis) contains a very convenient and readily available historical development of this chemistry, which is recommended for interested readers.

[304] J.-M. Basset. Yves Chauvin (1930–2015). *Nature.* 519:159 (1915).

Re, Os, Ir) which is known to form carbonyl and/or cyano complexes. Carbon monoxide (CO) fairly obviously is closely related to carbene (CH$_2$) and olefins (=CR$_2$). Pi-complexes of olefins with transitions metals (e.g., Zeise's salt) were well known.[305]

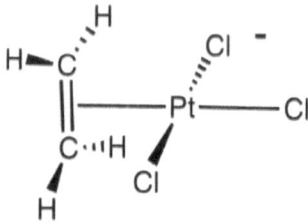

Zeise's salt   K[PtCl$_3$(C$_2$H$_4$)]·H$_2$O

Drawing by Smokefoot, source Wikimedia Commons

In these complexes the pi-electron density of the olefin is donated into a d-orbital of the metal and the carbon-carbon bond length increases. At some point, the remaining C-C sigma-bond of the ligand becomes unstable relative to the carbonyl-like structure. Thus, it is not hard to believe that what starts out as a simple pi-complex of an olefin with a transition metal (with lots of low level d-orbitals) can reversibly rearrange into a compound having a carbonyl-like bond:

---

[305] William Zeise (1789–1847).

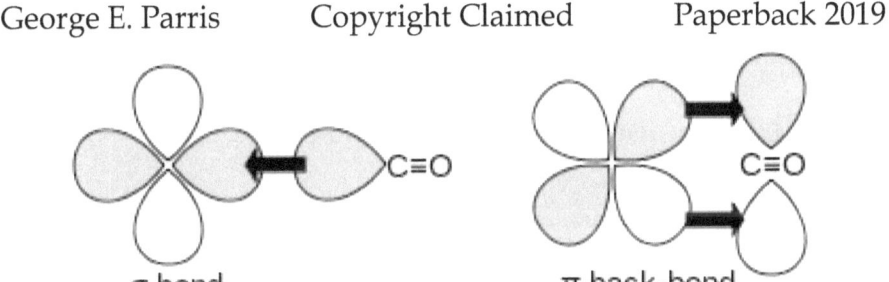

σ bond   π back-bond

Drawing by sarang, source Wikimedia Commons

The intermediate between these extremes was proposed by Chauvin to account for the metastasis of olefins:

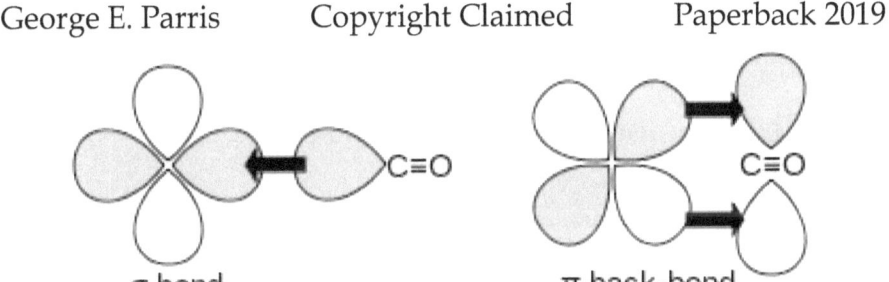

Drawing by V8rik, source Wikimedia Commons

Overall, the mechanism exchanges =CR$_2$ groups intact to produce the most stable olefin from a mixture of olefins. Chauvin's mechanism was challenged, but eventually it proved to be correct. Robert Grubbs (1942-) and Richard Schrock (1945-) have progressively refined the conditions of the reaction and optimized the catalysts (e.g., Grubbs' Catalysts) for synthetic applications on the industrial and laboratory scale. For their contributions, Chauvin, Grubbs and Schrock were awarded the Nobel Prize in Chemistry in 2005.

Palladium and Platinum Catalyzed Coupling Reactions

In 2010, the Nobel Prize in Chemistry was awarded to Akira Suzuki (1930-), Richard Heck (1931-) and Ei-ichi Negishi (1935-) for their work involving changes in the oxidation states of Pt and Pd complexes that can be useful to coupling organic radicals.

For example, a coupling reaction was originally observed using Pd(II) salts by Tsutomu Mizoroki (1971) and Heck (1972) and was based on earlier work on coupling ArX with olefins by Heck (1969). In these cases, the Pd(II) was initially reduced to Pd(0) by reaction with the olefin. In 1974, Heck greatly improved the reaction by pre-reducing the Pd(II) to bis(triphenyphosphine)Pd(0) by adding triphenylphosphine (Tpp) to the Pd(II) solution (triphenylphosphene oxide is the byproduct of this reaction).

The (Tpp)$_2$Pd(0) catalyst readily reacts with Ar-I to produce stable square-planar Ar-Pd(II)-X complexes. These are stabilized by the same p-orbital back-bonding described above. Initially, the complex is produced by cis-addition (Tpp$_2$Pd(II)ArX); but the labile complex rearranges to the more stable trans-configuration (TppArPd(II)TppX).

Then an olefin can form a pi-complex, which is immediately converted into a new sigma-bonded complex by shift of the Ar-ligand to the olefin (forming the most stable carbanion). In this case, however, the alkyl (R-) complex is unstable because of the hydrogen on the carbon and a weak base readily removes that proton to generate an olefin (by withdrawing the sigma-bond from Pd(II). The resulting complex is unstable and releases the anion (I-) to regenerated the Tpp2Pb(0) as the new olefin is released. In the presence of Ar-I and olefin, the reaction is, thus, catalytic.

The Hech Reaction

Negeshi and coworkers investigated similar coupling reactions involving organometallic (R-M) and R-X

$$R-X + R' \diagup\!\!\diagdown \xrightarrow[\substack{base \\ -HX}]{Pd^0} R' \diagup\!\!\diagdown\!\!\diagup R$$

Drawing by ~K, source Wikimedia Commons

compounds in the time period 1974-77.[306] And, Suzuki and co-workers reported similar reactions where the metal is boron in 1979.[307]

Asymmetric Hydrogenation

The principle that an optically active catalyst could produce optically active products was first demonstrated in 1966 in Japan by the group under Hitosi Nozaki (1922-) at Kyoto University.[308] The work was associated with a mechanistic study of the carbene

---

[306] E. Negishi, A.O. King, N. Okukado. Selective carbon-carbon bond formation via transition metal catalysis. 3. A highly selective synthesis of unsymmetrical biaryls and diarylmethanes by the nickel- or palladium-catalyzed reaction of aryl- and benzylzinc derivatives with aryl halides. *J. Org. Chem.* 42 (10): 1821–1823 (1977).

[307] N. Miyaura, K. Yamada, A. Suzuki. A new stereospecific cross-coupling by the palladium-catalyzed reaction of 1-alkenylboranes with 1-alkenyl or 1-alkynyl halides. *Tetrahedron Letters* 20 (36): 3437–3440 (1979).

[308] H. Nozaki, S. Moriuti, H. Takaya, R. Noyori, *Tetrahedron Lett.* 5239 (1966); and H. Nozaki, H. Takaya, S. Moriuti, R. Noyori, *Tetrahedron.* 24:3655 (1968).

reaction of ethyl diazoacetate with styrene.[309] One of the junior authors was Ryōji Noyori (1938-).

In the 1960s, G. Wilkinson, formerly of Harvard (1951-55), was now at Imperial College London focusing on inorganic complexes with catalytic potential. In particular, Wilkinson is perhaps best known for the "Wilkinson catalyst"[310] $RhCl(PPh_3)_3$. This catalyst was found to be effective at hydrogenation reactions. The accompanying figure (obtained from Wikimedia Commons) very nicely documents the changes in the valence electrons and oxidation states of the metal (Rh(I) and Rh(III)) in the catalytic process.[311]

---

[309] The reaction became industrially important in the manufacture of synthetic pyrethroids in the 1970s and beyond.

[310] Prepared by the reduction of Rh(III) chloride hydrate with triphenylphosphine.

[311] J.A. Osborn, F.H. Jardine, J.F. Young, G. Wilkinson. The preparation and properties of tris(triphenylphosphine)halogenorhodium(I) and some reactions thereof including catalytic homogeneous hydrogenation of olefins and acetylenes and their derivatives. *J. Chem. Soc.* A:1711–1732 (1966).

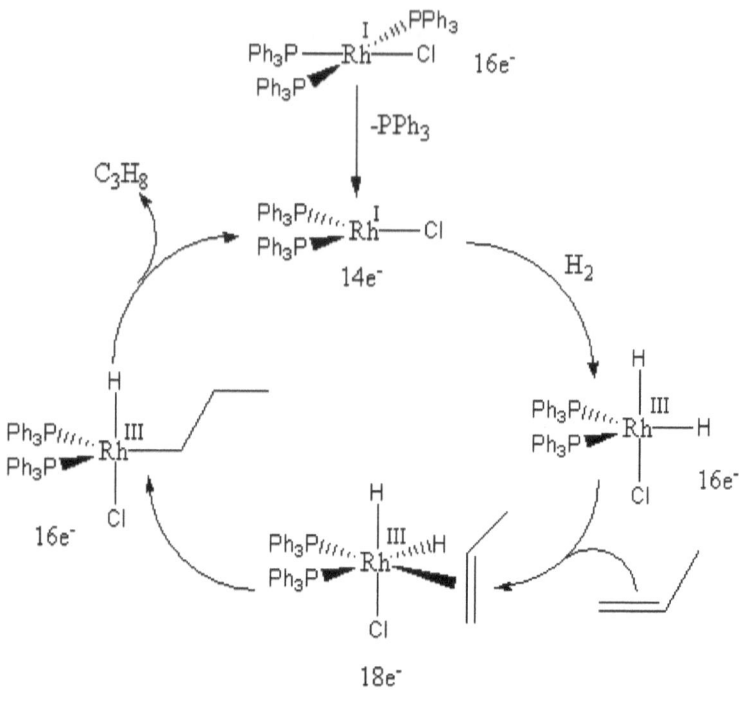

Catalytic hydrogenation of propylene

Drawing by Nanotube7, source Wikimedia Commons.

Meanwhile, William Knowles (1917–2012) was working for Monsanto and came to the frustrating realization that biological production of L-glutamate, L-lysine and L-menthol was much more expensive that using synthetic chemistry to make the racemic product, but resolution of the optical isomers was very expensive. The ideal solution to this problem would be to have a chiral catalyst that would do the necessary reactions to produce the desired optically active product directly.[312] He, thus, tried an obvious approach: He introduced optically active phosphines (prepared by the method of Mislow)[313] into the Wilkinson catalyst (about 1968).[314]

While Knowles was trying to produce a successful commercial process, Leopold Horner (1911–2005), who had been involved with synthesis of asymmetric phosphines using the Wittig reaction, had the same idea and reported asymmetric hydrogenation in the open literature.[315] This was probably a shock and inspiration for Knowles, but, Horner's interests carried him away from the hydrogenation reaction.

The initial experiments (independently by Horner and Knowles) with chiral methylpropylphenylphosphine proved the principle, although it was not economically

---

[312] Knowles' Nobel Lecture (2001).

[313] Kurt Mislow and Korpiun, *J. Am. Chem. Soc.*, 89, 4784 (1967).

[314] US Patent 3849480 A. Catalytic asymmetric hydrogenation. Priority date 9 Sep 1968.

[315] L. Horner, H. Siegel, H. Büthe. Asymmetric catalytic hydrogenation with an optically active phosphinerhodium complex in homogeneous solution. *Angew. Chem. Int. Ed. Engl.* 7(12):942 (1968).

useful. About this time, the medicinal utility (hence economic value) of L-DOPA became apparent and Knowles found that his system gave some degree of asymmetric synthesis. This justified a serious investment by his employer. His research team was amazed when on their 6th modification of the phosphine (methylcycohexyl-o-anisylphosphine) they were able to obtain 88% of the desired optical isomer. The laboratory discovery led quickly to a full-scale commercial process. Of course, successful chemical processes end up covered by patents and publication is often delayed.[316] The Knowles group continued their research with bidentate chiral diphosphines, which were successful (1971).

Meanwhile, back in academia, Ryōji Noyori (1938-) and Karl Barry Sharpless (1941-) passed through Harvard in the late 1960s.[317] Returning to Japan, Noyori thought that asymmetric hydrogenation would be

[316] B.D. Vineyard, W.S. Knowles, M.J. Sabacky, G.L. Bachman, D.J. Weinkauff. Asymmetric hydrogenation. Rhodium chiral bisphosphine catalyst. J. Amer. Chem. Soc. 99(18):5946–5952 (1977).

[317] Noyori was a post-doc (1969-70) working for E.J. Corey and Sharpless was a post-doc working for Konrad Bloch (1912-2000).

more generally useful and began following a path similar to Knowles. In particular, the success of the bidentate phosphine ligands inspired Noyori (1974) to begin work with (2,2'-bis(diphenylphosphino)-1,1'-binaphthyl), a.k.a., BINAP.

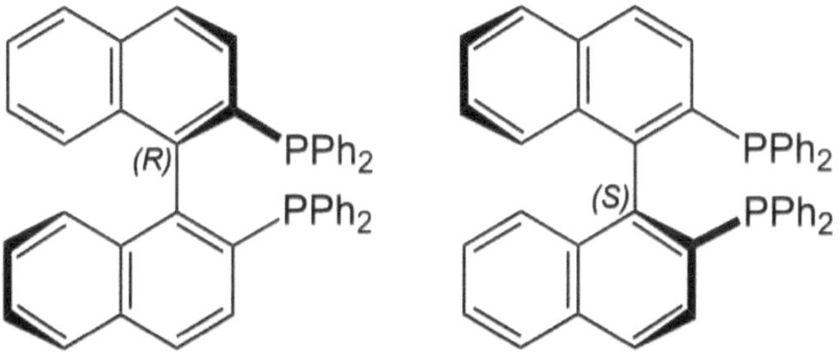

Drawings by Jü, source Wikimedia Commons.

However, his group was not able to reliably resolve the two isomers until 1978. The compound is attractive because rotation around the 1,1'-bond and the P-C bonds allows the ligand to firmly coordinate with ions of varying size.[318] His research group was finally successful in obtaining near 100% asymmetric reductions with this ligand in 1980. However, the oxidative addition of H_2 to the Rh(I) complex (see figure above) was not optimal and they found that by

---

[318] Noyori, Nobel Lecture December 8, 2001. This is a very nice document that summarizes the research clearly. There is also an excellent section on these catalysts in Wikipedia.

moving to ruthenium BINAP–Ru(II) they obtained better yields and faster rates of reaction (1986). The ruthenium(II) never changes oxidation state in the hydrogenation. Olefins and ketones also can be asymmetrically reduced.

Asymmetric Epoxidation

After his post-doc at Harvard, Karl Barry Sharpless (1941-) obtained a position at MIT. His world was almost immediately shattered by the loss of an eye in 1970 as the result of an exploding NMR tube. Fortunately, he overcame this setback. He had developed an interest in asymmetric epoxidation and followed closely developments in asymmetric catalytic reactions. He managed to obtain low levels of asymmetric oxidations in 1974, but did not consider the results worth publishing. In 1980, he returned to Stanford, where he had earned his PhD, and almost immediately published a breakthrough paper with Tsutomu Katsuki (1946-2014).[319] In 1990, groups lead by Eric Jacobsen (1960-) and Katsuki independently

---

[319] T. Katsuki , K.B. Sharpless. The first practical method for asymmetric epoxidation. *J. Am. Chem. Soc.* 102(18):5974–5976 (1980).

achieved success with asymmetric manganese catalysts in the preparation of epoxides.

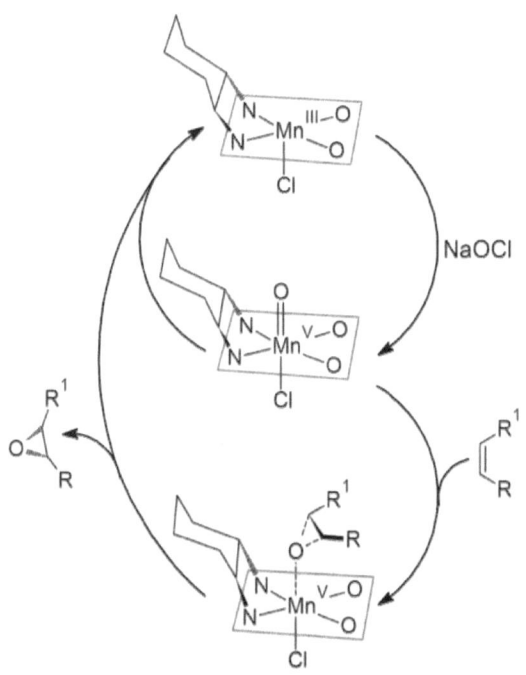

Drawing by LHcheM, source Wikimedia Commons

V. Post-Modern Organic Chemistry (1980- ?)

As we have seen in the development of transition metal catalysis and asymmetric synthesis described above, organic chemistry has continued to innovate into the 21st century. But, because of my personal experiences, I take a "post-modern" view of the events that have occurred since 1975. Clearly, there are overlaps in time of different movements that make my arbitrary divisions of period used in this book fuzzy and my transition from "modern" to "post-modern" can be criticized, not only for its demarcation (1980) but also as to whether it has happened at all.

My rather dark view of organic chemistry in what I choose to call the post-modern era is based on four principal issues: (1) Exhaustion of the "low hanging fruit" of chemical discovery involving molecules of molecular weights less than 5,000. (2) The failure of organic chemists to take the lead in the field now known as "molecular biology," which began with the elucidation of the structure of DNA (by a physicists and a biologists). (3) The environmental movement that has now completely encapsulated the profession of "chemist" in a network of regulations that ensure that chemists cannot be entrepreneurs in the way that they were pre-World War II. (4) And, the combined effects of (i) exponential growth in the number of PhD chemists, (ii) systematic importation of foreign talent into the United States, (iii) systematic preferences to women and minorities in an attempt to make up for generations of under representations in academia in a single generation. Overall, these effects (i-iii) make the profession of "chemist" (especially motivation for a career in academia) very difficult to achieve especially for "white males" who have traditionally led this field. In the following chapters, I will explore each of these forces that are shaping "organic chemistry" today.

1. The Impact of Environmental Regulations of the Chemical Industries

"The times.. they are a changin'"... Bob Dylan (1964)

I have begun this chapter that I call "post-modern organic chemistry" in 1980, because I earned my PhD in 1974 and in 1980 the rules and regulations radiating from the first major environmental laws targeted directly at the chemical industry (i.e., the Toxic Substances Control Act (TSCA) and the Resource Conservation and Recovery Act (RCRA)) [320] were being implemented by the US Environmental Protection Agency. The Comprehensive Environmental Response, Compensation and Liability Act (CERCLA) a.k.a. "Superfund" also indirectly affected chemical company operations because of the liability that was assigned to many of them for historical activities.

In the following sections, I will briefly summarize the environmental movement and the relevant environmental laws that directly affect organic chemistry and the careers of organic chemists. Sadly,

---

[320] TSCA is commonly pronounced ToS-CA. RCRA is commonly pronounced RiC-RA.

to my knowledge, no chemist is taught anything about these laws at any time during his/her academic training.

Indeed, all the training I received as an undergraduate and graduate student prepared me for one thing: academic research. There was never any discussion about the chemical industry, economics of chemistry, the politics of chemistry, or career opportunities in chemistry or career paths. Although a few of my mentors (both Long and Ashby) had worked in industry there was never any discussion of that as a career option. As one of the better chemistry majors, the inference was that my best approach would be to follow the R.B. Woodward model...devote myself to academic excellence and doors to an academic career would open. They never did.

Even when I was at the point of writing my thesis, I was never counselled regarding any career, Ashby never even tried to help me find a post-doc position. I found my own post-doc (at the NBS/now NIST a government agency). There were people working for Ashby when I arrived in Atlanta who were still working for him as post-docs when I left. When I started looking for a job, I came across a full-page photograph in one of the popular glossy periodicals (I think *Look* or *Life* magazine) which showed a chemistry PhD sitting in his laboratory and he had pasted rejection letters on the ceiling. The ceiling was covered. I don't believe I know any one from my undergraduate of graduate school experience who got a position at a four-year school. The situation was so bad that I drafted a "pre-rejection" letter and sent it to a friend from a hypothetical back-woods college. It started something like this:

> *We realize that you have not applied to our school for a teaching position, but if you ever dream of working here, please do **not** apply.*

The system then (and I think now) does not really prepare chemist for a career, it prepares them to be laboratory technicians. If you want something else, you need to get it on your own.

Some background will help:

Herman Muller and Cancer Risk Assessment

Herman Muller (1890–1967) is blamed/credited for the cancer risk assessment policy in the US by most people who have tried to track it down. But in truth, he said almost nothing about cancer. His Nobel Prize (1946) was for work he did on *mutations*[321] caused by x-rays. As a student of Thomas Hunt Morgan (1866–1945) whose work with fruit-fly genetics in the 1920s revolutionized biology and begged for the discovery of DNA, Muller was a disgruntled theorist whose name was not attached to papers where he may have shared ideas with his more hands-on colleagues. On his own, he began an experimental inquiry concerning the potential for inducing mutations through physical means. First, he tried heat and discovered that sub-lethal heating had no observable effect on mutation rate. Neither did the chemicals that he tried. Then he hit on the idea of using x-rays and found that x-rays did cause mutations and that the number of mutations was proportional to the dose of x-rays (apparently with no threshold). In his 1946 Nobel speech he stated:

---

[321] The definition of "mutation" and changed over time and still means different things to different people. To Muller, it meant a change in the phenotype that was attributable to changes in the genetic material (which the Morgan Group associated with DNA/chromosomes as early as 1925). Today, we distinguish between damage to DNA (i.e., genotoxicity) and mutation, which is the expression of DNA damage in *succeeding generations* (*of cells or populations*). It was not recognized until the 1970s that genetic damage is frequent, unavoidable and most often repaired at the cell level.

"Both earlier and later work by collaborators (Oliver, Hanson, etc.) showed definitely that the frequency of the gene mutations is directly and simply proportional to the dose of irradiation applied, and this despite the wave-length used, whether X- or gamma- or even beta-rays, and despite the timing of the irradiation. These facts have since been established with great exactitude and detail, more especially by Timoféeff and his co-workers. In our more recent work with Raychaudhuri (1939, 1940) these principles have been extended to total doses as low as 400 r, and rates as low as 0.01 r per minute, with gamma rays. They leave, we believe, no escape from the conclusion that there is no threshold dose, and that the individual mutations result from individual "hits", producing genetic effects in their immediate neighborhood. Whether these so-called "hits" are the individual ionizations, or may even be the activations that occur at lower energy levels, or whether, at the other end of the scale, they require the clustering of ionizations that occurs at the termini of electron tracks and of their side branches (as Lea and Fano point out might be the case), is as yet undecided. But in any case they are, even when microscopically considered, what we have termed "point mutations", as they involve only disturbances on an ultramicroscopically localized scale. And whether or not they are to occur at any particular point is entirely a matter of accident, using this term in the sense in which it is employed in the mathematics of statistics."

The point in his talk that was lost was the fact that he *only studied* high energy radiation. Few people appreciate that the energy contained in a single x-ray photon (E = hν) is thousands of times as much energy as is required to break a C-C, C-N, C-O or C-H bond and he specifically noted that the relevant mutations might be associated with "clustering of ionizations." He goes on to say:

> *"Naturally, other agents than photons which produce effects of this kind must also produce mutations, as has been shown by students and collaborators working under Altenburg in Houston for alpha rays (Ward, 1935) and for neutrons (Nagai and Lecher, 1937), and extended in regard to the quantitative relations concerned by Zimmer and others working with Timoféeff (1936, 1937, 1938), and by others. Moreover, as Altenburg (1930, 1935) showed, even the smaller quantum changes induced by ultraviolet exert this effect on the genes. They cause, however, only a relatively small amount of rearrangement of chromosome parts (Muller and Mackenzie, 1939) and, in fact, they also tend to inhibit such rearrangement, as Swanson (1944), followed by Kaufmann and Hollaender (1944 et seq.), has found. Since the effective ultraviolet hits are in the form of randomly scattered single-atom changes in the purines and pyrimidines of the chromosomes, rather than in groups of atom changes, it seems likely that clusters of*

ionizations are not necessary for the gene mutation effects, at any rate, although we cannot be sure of this until the relation of mutation frequency to dosage is better known for this agent."

But, you will note that in contrast to x-rays, UV rays *"also tend to inhibit such rearrangement."*

It is clear that his no-threshold argument only applies to high energy radiation. The reason for that is that (unknown at the time) living cells have developed a vast array of DNA repair mechanisms, which adequately deal with *isolated* single- and double-strand breaks in DNA. Indeed, since many DNA-damaging events have nothing to do with the (UV) radiation, the induction of DNA-repair proteins by exposure to UV actually reduces the net number of mutations passed forward.

But, some of the massive rearrangements caused by high-energy "hits" (i.e., multiple instantaneous DNA breaks followed by random re-combinations; now known as *chromothripsis*) are beyond repair (on any timescale) and/or the repair mechanisms are swamped by the numerous instantaneous strand-breaking events associated with high energy radiation (e.g., decay of ^{40}K ions associated with phosphate anions of DNA).

Keep in mind that Muller was almost exclusively concerned with effects on populations and said very little about cancer. The association of his work with cancer and cancer risk assessment did not come to the front until the 1960s when the idea that "mutations (i.e., genotoxic events) cause cancer" was put forth and the *linear no-threshold dose-response model* proposed by Muller for *mutations* was applied to cancer.[322]

The role of decay of naturally-occurring radioactive potassium-40 has been largely ignored in the context of cancer risk assessment. Based on the data by Moore and Sastry (1982)[323], there are approximately 10^{-8} decays per day per cell. This result may sound minor and you may feel safe; but when you realize that there are approximately 30 trillion ($\sim 10^{13}$) cells in the human body with most major organs having at least a billion cells, it becomes clear that ^{40}K decay is occurring in ever organ every day. So, in each major organ (skin, liver or

---

[322] I have discussed this in detail in my book *The Myth of the Linear, No-Threshold Dose-Response Relationship for Carcinogens* self-published on Amazon/Kindle (2013).

[323] Moore FD, Sastry KS. Intracellular potassium: 40K as a primordial gene irradiator. *Proc Natl Acad Sci U S A.* 79(11):3556-9 (1982).

lungs) there are probably about ten ^{40}K disintegrations every day of your life…and yet most people do not die of cancer. More about this in a moment.

Silent Spring (1962)

Rachael Carson (1907–1964) died of complications from breast cancer at the age of 57. She was a marine biologist and gifted writer who won awards for several books published in the 1950s. Thus, people took notice of what she wrote. And, starting in 1957 she developed an interest and concern about the use of pesticides, especially DDT. She argued that DDT poisoned non-target wildlife (especially birds that normally ate insects) and the title of her book Silent Spring (1962) suggested the extinction of bird species that chirp loudly on spring mornings. The book was an immediate, popular success and provoked President John F. Kennedy to ask his Science Advisory Committee to investigate her claims. The committee generally supported Carson's concerns and recommended phase out of DDT and other persistent toxic pesticides in May 1963.

Her concern was reasonable considering the unmitigated faith that had been placed in pesticides after WWII. In particular, it was clear that DDT had saved millions of lives and rid the US and southern Europe of malaria. Insect pest in crops were being subdued and the general attitude regarding chemistry (the source of numerous new plastics and fibers) and pesticides in particular was most favorable in the late 1950s. Isaac Asimov (1920–1992) had recently set many school boys on to chemistry with his very readable books e.g., the *World of Carbon* (1958). I was one of those boys and I was excited to find that I was in good company when I read the following review by W. Nugent posted on amazon.com on March 21, 2007:

> *"This is the book that started it all for me. I went on to a fulfilling 30-year career in organic chemistry* [eventually becoming the Chair of the Division of Organic Chemistry of the American Chemical Society]. *If you are looking for a book to open the mind of a child or grandchild to an amazing and essential field of science, this is the one. The companion book "The World of Nitrogen"* [also out of print] *is also very engaging. Since Dr. Asimov passed on several years ago, I post my note of thanks here in hopes they have high-speed access in the Great Beyond."*

So, Dr. Carson gave us a warning. Most chemists at the time, though she overstated her case and suspected that her disease influenced her objectivity. But, it was well received by the general public and triggered major anxiety among that fraction of the population that is obsessive compulsive and chemophobic. More importantly, the broad-based "conservation movement" began to focus on cancer and chemicals.

The 1960s were a period of social upheaval in the US. The threat of nuclear war with the Soviets was real. A popular president was assassinated. Through political miscalculation and incompetence the war in Vietnam was made unwinnable and unstoppable. The political sensitivities of the Johnson Administration (1963-1968) activated the Civil Rights movement. More emotional and social revolution occurred than any time since the American Civil War. Out of this came a commitment to embrace a more socially conscious national ethic. For example, William Joseph Sparks (1905-1976) president of the American Chemical Society (1966) had this to say:

> *"The scientific profession has become much larger than medicine, law or the clergy. Yet, many young scientists are not taught by their professors to feel an obligation to society in their work."*

Nothing much happened with DDT until 1967 when the Environmental Defense Fund (EDF) was formed as a private lobbying group to force action of DDT. They filed law suits and lobbied against the use of DDT; and, starting in 1969, they had the benefit of work by David Peakall (1931–2001) who showed an inverse correlation between raptor egg-shell thickness and DDT exposure in some endangered bird species.

The Environmental Protection Agency

In January 1970, the National Environmental Policy Act (NEPA) was passed by Congress. President Richard Nixon formed the US Environmental Protection Agency (December 1970) to administer NEPA[324] by drawing together bits and pieces of existing Federal Agencies. Nixon appointed William D. Ruckelshaus (1932-) an attorney to be the administrator of the new agency.

The responsibility for DDT (Rachel Carson's primary target) had been transferred to EPA along with the enforcement elements of the Department of Agriculture. Thus, EDF filed suit against the pesticide registration of DDT by EPA in 1971 and won a court

---

[324] Pronounced Nee-Pa.

ruling directing EPA to review DDT's registration with an eye to ending its pesticide registration.

After review, EPA (based on analysis by the staff inherited from the Department of Agriculture) recommended against de-registration. But, EDF successfully organized a public rejection of this decision by arguing that the EPA staff was biased to favor agricultural businesses. This prompted a further review with a public hearing 1971–1972. The decision by Hearing Examiner Edmund Sweeney was as follows:

> *"DDT is not a carcinogenic, mutagenic, or teratogenic hazard to man. The uses under regulations involved here do not have a deleterious effect on fresh water fish, estuarine organisms, wild birds, or other wildlife. The evidence in this proceeding supports the conclusion that there is a present need for essential uses of DDT."*

Regardless, politics enters into virtually everything the USEPA does. In the summer of 1972, events were spinning out of control for the Nixon Administration: Of course, it was an election year (November 1972). On June 17th, five burglars including a Republican security aid[325] were arrested during a botched break-in at the

---

[325] James W. McCord was the security director for the Committee for the Re-election of the President.

Democratic National Headquarters at the Watergate
Hotel in Washington, DC. The press, that was already
less than sympathetic to Nixon, immediately made this
into the story that would not go away. Discussion of
the Watergate episode is far beyond our scope, but the
adverse publicity at a sensitive time undoubtedly went
out through the entire Nixon Administration, including
his EPA Administrator (who was by no means an
environmental activist). This may explain why, in spite
of the ruling of the hearing officer, Ruckelshaus
attempted to strike a middle ground by issuing a
partial ban of DDT. Instead of pleasing everyone, he
offended everyone. The manufacturers (with strong
economic interest) and the EDF (realizing that public
momentum was on their side and the Administration
was compliant) sued EPA over the decision. In 1973,
the court essentially sustained the ban of DDT with
public health exceptions (e.g., in case of an outbreak of
malaria) as laid out by Ruckelshaus.

DDT (principal component)

Drawing by Leyo, source Wikimedia Commons

Nixon's troubles continued to spiral. Thus, environmental lobbyists and his opposition in Congress were able to obtain more or less anything they wanted. The Endangered Species Act was passed in December 28, 1973. And, Richard M. Nixon resigned August 9, 1974. He left behind and unelected president, Gerald Ford Jr., who was portrayed as honest but bungling. Ford would be up for election in 1976.

The Toxic Substances Control Act (PCBs)

EDF's success in banning DDT set a pattern. DDT was not the only chemical used widely in the environment; what about all those others?   It was not hard to put together a politically significant coalition that the Ford Administration needed to take into consideration. Of course, by now the Congress (in the wake of the Watergate scandal) had swung strongly to the Democratic Party. Interest in a comprehensive law to regulate industrial chemicals as well as pesticides was growing. The news media and lobbyists began the habit of bringing up some story of adverse effects of some chemical or other about every month.

John R. Quarles Jr. (1935-2012) was a Harvard-educated environmental lawyer who came to the EPA from the Department of the Interior and became acting Deputy Administrator for the EPA under William D. Ruckelshaus on April 30, 1973.

About that time, a case almost made to order to make the environmentalists' point unfolded in Michigan. Michigan has brine wells that produce bromide containing salts and hence it became a center for the manufacture of bromine-containing chemicals. One of the products that was typical of the industrial mentality of the day was polybrominated biphenyl (PBB). The process is disgustingly devoid of elegant chemistry from an organic chemist's perspective. Biphenyl is simply brominated (electrophilic aromatic substitution) until a certain weight percent of bromine is retained in the product (PBB). The product is characterized only by its gross composition and melting range. Cheap to make, and effective as a flame retardant (combined with antimony trioxide) in thermoplastics (e.g., television cabinets, which at the time were quite bulky and carried lots of vacuum tubes).

Magnesium salts were also a byproduct of bromine recovery. The company mixed the PBB with an anti-caking agent and sold it under a trade name in paper sacks. As the story has been handed down to me, at some point the plant ran out of the pre-stenciled bags for PBBs and the workers simply borrowed bags from the line that produced magnesium salts that was used as a nutrient in animal feed. Thus, the PBB (many pounds of it) was shipped to feed mills in Michigan and surrounding states where equally ignorant workers dutifully mixed it with other components and packaged it as animal feed, most of which went to Michigan dairy farms in 1973.

Within a few months animals were dying and showing a variety of symptoms including abnormal skin, hair and hoofs. The dairy farmers noticed similar effects on their families. The source of the problem was not understood until April 1974.[326] Interestingly, the population of exposed humans has been studied continuously through 2014 and any effects on health have been hard to pin down (e.g., non-statistically significant increases in cancer). The irony here is that because this episode was clearly and accident and mainly managed by the Food and Drug Administration (FDA) and local health officials, it was treated as a local

---

[326] Fries GF. The PBB episode in Michigan: an overall appraisal. *Crit Rev Toxicol*. 16(2):105-56 (1985).

issue and did not really seem to make it onto the EPA/EDF radar. They needed a bigger fish to fry.

Since I am relating some of this story from a first person perspective, it should be noted that I started a National Research Council post-doc at the Bureau of Standards (now NIST) in Gaithersburg, MD in November 1974. I had written a proposal to investigate environmental fate of organo-antimony compounds based on my undergraduate research at NC State University. This was a two-year appointment and I was well-liked at NBS/NIST; but when January 1976 rolled around and there was no opening in the area in which I was working, it became urgent that I find a new job. The only tenure-track academic position that I was ever interviewed for was at Auburn University in 1975; obviously I did not get the job. My mentors at NIST tried to get me interested in moving into the inorganic glass area (e.g., making IR-transparent glasses for guided missiles), but it was clearly not my cup of tea. Then one day, they came to me and said that the EPA Office of Toxic Substances wanted a chemist to help in their monitoring program...I would be the object of an inter-agency agreement (I officially worked for NBS, but the money came from EPA).

While I was working on arsenic chemistry at NBS/NIST, Tom Kopp and others at the EPA were collaborating with Congressional staff to create a bill that would create broad new powers for the EPA in regulating all industrial chemicals (not merely drugs, food additives and pesticides) and provide a safety net to respond to any chemical hazard. The bill was known as the Toxic Substances Control Act and a small Office of Toxic Substances was formed to shepherd it through Congress. Apparently, Tom Kopp identified polychlorobiphenys (PCBs) as the ideal candidate to be the TSCA "poster child." These materials were produced in huge quantities and used mainly as a non-flammable fluid in large electrical transformer and as a heat transfer fluid. There was a family of products manufactured mainly by Dow Chemical that had various viscosities determined mainly by the percentage of chlorine contained. The liquid was regarded as very low toxicity and was used and disposed carelessly, but unlike its hydrocarbon relatives, it was not readily biodegraded and it readily partitioned into fat which carried it up the food chain (like DDT).

TSCA was (and is) quite unlike other Federal statutes that regulate chemical substances in that it is inclusive rather than exclusive. Rather than putting the burden on the government to prove that a substance should be regulated, it starts with the proposition that all chemicals are hazardous until proven safe in some particular use. The reach of TSCA was so broad that specific areas had been defined as exempted from its scope (e.g., things regulated under the Atomic Energy Act and tobacco). When I was first involved with the USEPA (early 1976), there was speculation among my colleagues that an entirely new agency might be formed just to administer TSCA and there was a brief period when the Office of Toxic Substances considered regulating genetically altered organisms.[327]

The only real limitation built into TSCA was in Section 9, which directs the EPA to defer regulation to other federal laws where they are applicable. But at the time of passage of TSCA, neither RCRA (the Resources Conservation and Recovery Act…read *management of chemical waste*) nor CERCLA (the Comprehensive Environmental Response Compensation and Liability Act…a.k.a. Superfund…read *pre-existing sites of chemical*

---

[327] The Second Asilomar Conference was held in 1975 and there was a major public meeting in Washington, DC in 1976, which I attended on behalf of the USEPA Office of Toxic Substances. Fortunately, NIH took the lead on this area. Some of the broad regulations suggested by naïve EPA bureaucrats would have made heterosexual sex illegal.

contamination) had been passed. Thus, in principle, these activities could have been developed as regulations under TSCA.

The effect of TSCA on organic chemistry was to forever and completely make chemical entrepreneurship illegal. Most of the chemists that we have written about in the 1800s and early 1900s would not be able to legally launch a business based on a laboratory invention or discovery today. I even watched debate on the floor of Congress needed to exempt academic and commercial research chemicals from TSCA. An exemption is allowed under section 5(h)(3) for manufacture of chemicals (with no application to EPA) if the chemical is manufactured (synthesized):

> *"only in small quantities solely for the purposes of scientific experimentation or analysis, or chemical research on, or analysis of such substance, or another substance, including such research or analysis for the development of a product."*

Congress realized that without this exemption, chemical research and innovation of any kind would have moved out of this country and I would be typing this in German, French, Russian or Chinese.

At the time I reached EPA-OTS (on detail from NBS) the office had about thirty people and I think only three of us were chemist. Most of the employees had backgrounds in industrial hygiene, biology, toxicology, etc. My job was to assist the Monitoring Group find chemicals that would support the passage of TSCA and its implementation. This was all something of a culture shock for me because the motivation was completely political. There was an underlying bias against chemical manufactures among most of the employees and a general interest in empire building among the senior staff.

Actually, I did not realize what a career "gold mine" I had stepped into. In my carpool from Gaithersburg to Waterside Mall (401 M St SW) in DC, I was accompanied by Frank Kover (who became a Division director) and Charlie Auer (who became the head of the entire Office of Toxic Substances before he retired). Even the administrative assistant for our monitoring group earned a master's degree in biostatistics and ended up as a branch chief.

Ironically, the political appointees were from a Republican background. In particular, Glen Schweitzer was the head of the office. The law was passed in September 1976 and signed by Jerald Ford, just ahead of the elections; but it did not save him from losing to Jimmy Carter (from Georgia), who put together a coalition of Southern conservatives and New England liberals to win. Soon things changed at EPA. Schweitzer disappeared.

There was a push at EPA/OTS to find the next PCB-like problem to cement the role of TSCA and justify a massive scale up of the office. I did not really understand all the motives. I was generally uneasy with the facile use of data by EPA and the antagonistic relationship to industry. I was given the task of designing a program to monitor PBBs around production facilities mainly in New Jersey. In particular, human exposure was considered to be most important (i.e., most persuasive to the public). I hit on the idea of sampling hair from local barber shops (which could be classified by economic class, race and sex). Thus, in addition to air, soil, water and biota, we anonymously collected hair. Sure enough we found traces of PBBs in human hair near the production facilities. It was just what the agency wanted and I did not realize that I was near superstar status when I gave a press briefing (June 17, 1977).

Unfortunately, the contract with the people doing that work ended and the next effort was shifted to a much less competent group who I really did not know (I was not the project officer or the contract officer; I was just a technical assistant to my branch chief). In late June 1977, I was directed to expand the PBB studies by looking for PBBs in fish from major rivers. I contacted various state and federal fish and game agencies and arranged to have catfish from the Ohio River sent to the new contractor for analysis for PBBs. There was some urgency for this, which I did not understand. In late July, the urgency turned into an emergency and I found myself asking the contractor for data. The contractor had told me (on the phone) that they had found high levels of PBBs in Ohio River catfish. This information I passed to my branch chief and I was immediately told to get written materials sent to me for confirmation. I asked to have the data sent to me by air express (this was before Federal Express; it was literally placed on a passenger airplane and I picked it up at the airport). The chromatograms were not very definitive; just some rather wide irregular peaks. The contractor's method of confirmation was to monitor three ions by GC-MS that were selected because they were typical of PBBs. (Multiple ion monitoring had been selected to increase sensitivity.) They had peaks on all three ion channels at the same time and felt that they had confirmed the presence of PBBs.

I did not view their work as conclusive and I was very disturbed because the fish in question were many miles from any known source of PBBs and they were claiming that the fish had levels of about 10 ppm. I warned my boss via written memo on August 1, 1977 as follows:

> *I personally will not stand behind this data until it is confirmed by full scan GC-MS and I would like to have the retention time of the three (commercially available) isomers of monobromobiphenyl compared to the peaks observed in these samples. I recommend that these data not be "advertised" until confirmation is completed.*

Had the contractor been competent, full scan (all m/e values) mass spectra of the alleged PBBs peaks would have been promptly forthcoming. But, they were not. In the meantime, unknown to me, (most likely) Tom Kopp had been in contact with the congressional staff and told them that we had found high levels of PBBs in Ohio River catfish. My boss's reaction to my memo was to insist that I confirm the findings. I asked the contractor for more data, but they seemed confused or recalcitrant.

I found out what the urgency was about on August 3rd. I was asked to attend a hearing by senior staff of EPA OTS led by Gene Wallen (acting director) of the Office before the House Committee on Oversight and Investigation.

At the hearing, Wallen, Kopp and my office-mate Joseph Seifter were sworn to testify. Wallen provided about thirty minutes on the background of TSCA and our findings of PBBs in NJ. He never mentioned catfish. When he had finished (what seemed to me like a routine and rather boring presentation), he was questioned by the committee members. Mr. Luken began a cross-examination of Wallen on the topic of what constituted an "imminent hazard" (requiring extraordinary actions) under TSCA. Then unexpectedly, the bright camera lights came up and Luken asked about the findings of PBBs in catfish near Cincinnati, OH. I had never discussed this matter with Dr. Wallen (a biologist by training as I recall) and he tried to respond to the questions. In particular, he took the position that the identification of PBB was confirmed and that we only needed to tie down the concentrations.

Now, it was obvious why I was being pressured to confirm the findings. The pressure continued for about another week and when the full scan mass spectra finally arrived, they looked nothing like the spectrum expected for PBBs....they represented a slice through a chromatogram of fish fat that had responses on *almost every* ion channel (including the three that were originally monitored).

In the meantime, the story got into the *Washington Post* and I had a meeting with the FDA. I discovered real chemist doing real science there (at the time 200 C Street SW). Indeed, they had frequently monitored fish from that region and occasionally found a variety of pesticides and PCBs, by never PBBs. I ended up writing memos on August 5th and August 10th under pressure to "confirm" the findings. I finally had to tell my boss that if he wanted to confirm the findings he would have to write his own memo.

Steven D. Jellinek (Democratic appointee) appeared in the fall of 1977 to lead the OTS and I found a job at FDA. Unfortunately, I found FDA to be as boring as EPA was political. I left the government (never to return) in 1981, and went to work for a consulting firm in Rockville, MD where I found what it is like in the private sector where profits are critical to job security.

I found myself in a very interesting position. Between 1981 and 1988, the consulting firm had a range of contracts across the EPA (some of which I helped win) and I was able to work on projects for the TSCA Interagency Testing Committee, and several different divisions of OTS (including test rules development program, the pre-manufacturing notice review group); the RCRA listing program, the RCRA land disposal restrictions program; plastic waste report to Congress, etc. This was actually and amazing educational opportunity to learn about the chemical industry and the way that the EPA was regulating industrial chemicals (from the pre-manufacturing notice to the land disposal of waste).

TSCA from the Outside Looking in

Following passage of RCRA and CERCLA, TSCA is primarily involved with research, development, production and marketing of chemical products. As it stands, TSCA divides the world of chemicals into two groups: *Old chemicals* (i.e., chemicals acknowledged to be in commerce by various industries at the time the

TSCA Inventory[328] was created late 1970s) and *new chemicals*, chemicals not on the TSCA inventory. When a manufacturer, decides to move a chemical from "research" to "production" (manufacture), they are required to file a Pre-manufacturing Notice (PMN) with the EPA for the new chemical. The EPA reviews these data, which include the intended market, and evaluates the likely level of production (this is an economic analysis that compares the cost of production to existing markets and other possible markets) and the likely hazard (toxicity, environmental fate, etc.). The available toxicity data will be reviewed to determine if additional testing should be required. If so, the EPA can request a variety of tests to be performed before

---

[328] Note that to the bureaucrats "chemical" means "some commercial product or intermediate" as identified by a CAS (Chemical Abstract Service) number. Thus, "benzene" had a CAS number, benzene might also be a major component in some petroleum fraction (with a separate CAS number) and some other products (with still other CAS numbers). A typical list of "chemicals" might include "benzene, mercury, and PCBs" where benzene was a pure compound, "mercury" included all mercury compounds and "PCBs" included various mixture of many isomers of many related compounds (containing 1 to 10 chlorine atoms attached to biphenyl). Although the bureaucrats frequently talk of "concentration" they really mean "regulatory level" as *operationally defined* by some standard method of analysis.

they make a decision as to whether or not the chemical can be allowed onto the TSCA inventory (which contains about 60,000 items). EPA may also limit the market if they desire, requiring a "significant new use review" (SNUR) similar to a PMN if the manufacturer wants a product to move beyond the market niche to which it was initially restricted.

Under TSCA, an "Interagency Testing Committee" (ITC) has been established to review chemicals in commerce (i.e., on the TSCA inventory when TCSA was enacted) looking for chemicals with high exposure potential and insufficient toxicological data to evaluate its hazard. For example, in the early 1980s, bisphenol A[329] (a major component in epoxy resins and polycarbonates since the 1930s) was identified by the ITC as such a chemical.[330]

The chemical was referred to the Test Rule Development Branch and (as a contractor to EPA) I was the principal author on the Test Rule Support

---

[329] Condensation of phenol and acetone produces BPA and various other isomers and byproducts.

[330] Sarah Ann Vogel. *Is it Safe?: BPA and the Struggle to Define the Safety of Chemicals*. University of California Press. 2013 (pp.88-89).

Document for Bisphenol A (1984). In that document, we noted that some estrogenic effects had been reported for bisphenol A (BPA).

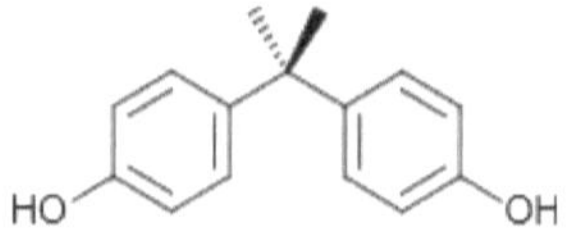

Bisphenol A

Drawing by Darkness3560, source Wikimedia Commons

In the report I included a drawing comparing BPA to estriol and estradiol:

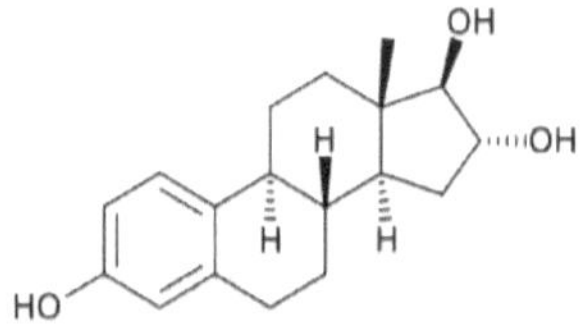

Estradiol

Drawing by Ayacop, source Wikimedia Commons

I do not believe EPA considered this very important at the time. However, twelve years later (1996), the publication of *Our Stolen Futures*[331] (with a forward by sitting Vice President Al Gore) got the EPA's attention and argued that BPA and other industrial chemicals were having subtle effects on the endocrine system. This concept has been controversial because of its non-

[331] Theo Colborn, Dianne Dumanoski, John Peterson Myers. *Our Stolen Futures*. Forward by Al Gore.

linear dose response claims and the political environment in which it was introduced.[332]

As a result of this controversy, there has been revived interest in bisphenol chemistry primarily among environmental scientists. Toxicologists have isolated a "metabolite" of BPA that they have found to be 100 to 1000 times more effective at binding to estrogen receptors than BPA.[333] This alleged metabolite is actually a likely contaminant in commercial BPA:

4-Methyl-2,4-bis(4-hydroxyphenyl)pent-1-ene (MBP)[334]

Drawing by Edgar181, source Wikimedia Commons

---

[332]Kaiser J. Endocrine Disrupters. Controversy continues after panel rules on bisphenol A. *Science*. 317(5840):884-5 (2007).

[333] Scott Lafee. BPA's Real Threat May Be After It Has Metabolized. UC San Diego. UC San Diego News Center. October 04, 2012.

[334] Imagine the olefin adding to the meta-position of the opposite ring to close a 6-membered ring. For example, 1,1,3-Trimethyl-3-(4-hydroxphenyJ)-S-indanol.

I cannot imagine an organic chemist looking at this molecule and not hypothesizing that it is a byproduct of synthesis of BPA (not a metabolite). For example, Dianin's[335] compound (4-p-hydroxyphenyl-2,2,4-trimethylchroman) is a significant component of commercial BPA.[336]

4-p-hydroxyphenyl-2,2,4-trimethylchroman

Drawing by Edgar181, source Wikimedia Commons

In a different study for the Economics and Technology Division of OTS, we were tasked with evaluating the exposure of people to epichlorohydrin and determining if epichlorohydrin could be eliminated from commerce.

---

[335] Alexander Pavlovich Dianin (1851–1918) discovered bisphenol A in 1914.

[336] M. Terasaki et al. Estrogenic activity of impurities in industrial grade bisphenol A. *Environ Sci Technol.* 39(10):3703-7 (2005).

Epichlorohydrin is used as an intermediate (i.e., no end uses). As lead author on this report (1984), I and other chemists and chemical engineers looked into alternative pathways to the principal use of epichlorohydrin in epoxy resins. We concluded that there were two alternative pathways to manufacture the same end product (i.e., the diglycidyl ether of bisphenol A) and examined the cost of production and waste generated in each case. Based on 1983 cost data, we considered the peracid epoxidation of the diallyl ether of bisphenol A to be prohibitively expensive. But, over the last thirty years several new methods of epoxidation have come forward that would greatly improve the economics and perhaps reduce the byproduct waste generation (see below). Nonetheless, to my knowledge, EPA has taken no action on epichlorohydrin. The following figures are from the 1984 report:

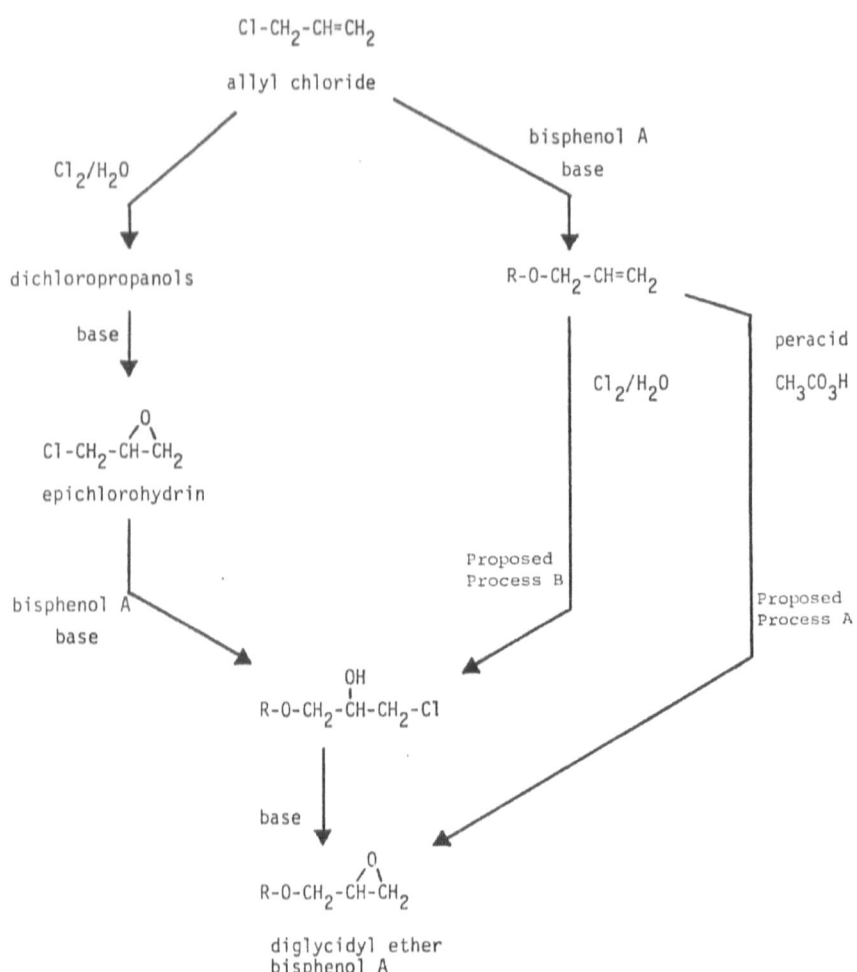

Table 4.2.3A. Raw Materials Needed to Produce One Pound of DGEBPA

	Cost[a] $/pound	Conventional Process pounds[b]	Conventional Process cost($)	Proposed Process A pounds[b]	Proposed Process A cost($)	Proposed Process B pounds[b]	Proposed Process B cost($)
Allyl chloride (FW 75)	0.61	0.82	0.50	0.61	0.37	1.0	0.61
Chlorine (FW 71)	0.08	0.77	0.06	--	--	0.53	0.04
Caustic (50% solu.) (FW 40)	0.15	0.87	0.13	--	--	0.79	0.12
Bisphenol A (FW 228)	0.67	0.74	0.50	0.93	0.62	0.82	0.55
Peracetic acid (35%) (FW 76)	2.66	--	--	0.61	1.63	--	--
Potassium carbonate (FW 138)	0.34	--	--	2.29	0.78	--	--
Acetone	0.21	--	--	0.15	0.03	--	--
Methylene chloride	0.29	--	--	0.15	0.04	--	--
Methyl Isobutyl Ketone	0.47	0.09	0.04	--	--	0.09	0.04
DGEBPA product:		1.0 pound/$1.23		1.0 pound/$3.47		1.0 pound/$1.36	
Solid wastes produced:[c]		2.29 pound		3.74 pounds[d]		2.23 pounds	

[a]Based on current prices for large volume purchases.
 Data for peracetic acid are from FMC. Other data are from CMR (Hammaker, 1984).
[b]Based on Dynamac chemical and engineering estimates.
[c]Disposal cost are not considered here but could be about $0.05/lb.
[d]About one pound of this is CO_2 vented from the reactor.

Figure and Table from the Report. Cost figures were for 1983-84.

RCRA

The Resource Conservation and Recovery Act (RCRA) was the first new legislation to supersede the authority of TSCA. And, it quickly became one of the primary examples of EPA bureaucratic overreach.

The Act defines "hazardous waste" as

> "a solid waste . . . which because of its quantity, concentration, or physical, chemical, or infectious characteristics may (A) cause, or significantly contribute to an increase in the mortality or an increase in serious irreversible, or incapacitating reversible, illness; or (B) pose a substantial present or potential hazard to human health or the environment when improperly treated, stored, transported, or disposed of, or otherwise managed" (USC 42 § 6903(5))

I present this definition here because many of the wastes being regulated under the Act as "listed hazardous waste" simply do not meet this criterion.

Implementation of this law was left to the USEPA. The language of the Act instructs EPA to

> "...promulgate regulations identifying the characteristics of hazardous waste, and listing

particular hazardous wastes (within the meaning of section 6903(5) of this title)…" (USC 42 § 6921(b)(1))

This language suggests to me that EPA should (a) develop defining characteristics of "hazardous waste" and (b) lists those wastes that *meet those characteristics*. But, EPA took an entirely different approach:

> The USEPA (A) defined characteristics to identify "hazardous waste," and (B) where gaps in the characteristics might exist (in EPA's opinion), certain waste streams have been identified by rule as "hazardous waste."

Whether it intended to do so or not, EPA has created a dual structure for identifying hazardous waste (40 CFR 261.10 *characteristically* hazardous wastes & 40 CFR 261.11 *listed* hazardous waste). In my view, this approach has no basis in the mandate provided by Congress and greatly overreaches the objectives of the Act. Specifically, *listed hazardous wastes are not necessarily characteristically hazardous*; in fact, they generally are not characteristically hazardous.

The first rules appeared in 1980 as the EPA program became effective. Consistent with the directions from

Congress, the first method devised to identify
hazardous waste was to set up objective tests that
measured chemical, physical and toxicological
characteristics that were linked to the ability of the
waste to have the adverse impacts that Congress
wanted to protect against (40 CFR 261.20 - .24).[337] If a
solid waste fails any of the characteristic tests, it is a
hazardous waste, except for some solid wastes that are
excluded from being hazardous waste for various
technical reasons (40 CFR 261.4(b)).   This approach
makes logical scientific sense. The logic of the tests

---

[337] One of these tests was originally (1980) an Extraction
Procedure designed to simulate what would happen if a waste
were managed in an ordinary solid waste (municipal) landfill
(i.e., as a non-hazardous waste). In this test procedure, the
concentration of a few constituents for which Federal drinking
water standards (i.e., Maximum Contaminant Levels for public
drinking water, MCLs) had been developed (under the Safe
Drinking Water Act) were measured in the extract. The RCRA
standards were obtained by assuming there would be some
dilution or attenuation of the leachate before it could become
drinking water and the Extraction Procedure for
characteristically identifying hazardous waste used multiples of
the drinking water standards (e.g., 100 x the MCLs) as the
hazardous waste criteria. This entire concept was extended in
the mid-1980s with the Toxicity Characteristic rule in which a
longer list of constituents of concern was prepared and a more
elaborate leaching test (the Toxicity Characteristic Leaching
Procedure, 40 CFR 261 Appendix II) was developed.

follows from what happens in the real world and you can objectively decide if a non-excluded solid waste is hazardous *without a waste manifest or an attorney*.

In general, objectively measurable characteristics are an attractive way to classify hazardous waste, not only because you know when you enter the waste management system (40 CFR 261.3(b)), but also you know when you exit (40 CFR 261.3(d)(1)) *without filing any reports, submitting petitions or waiting for government decisions or rulemakings*.

Unfortunately, the characteristic approach had two weaknesses: (1) The original Extraction Procedure covered very few constituents and even the newer Toxicity Characteristic rule does not cover the entire universe of possible hazardous constituents, and (2) the characteristic is potentially susceptible to sham treatment (typically dilution). The EPA thought it would solve both these problems by providing lists of certain waste streams that the EPA has typically found to be hazardous through testing similar to the characteristic tests. In this concept, EPA does all the testing and sustains the burden of proof to demonstrate through the rulemaking procedure (including a regulatory impact analysis to identify the financial impact of the cost of regulation, EO 12291) that a waste is a hazardous waste. This sounds good to industry, and the EPA accomplishes its objective by defining the method of generation (an operational definition of the hazardous waste) and declaring that the waste stream must be tracked through manifest and managed as a hazardous waste without regard to normal variations in the exact constituents or their concentration.

EPA could have completed the job in 1980 at this point by promulgating a rule saying the *listed wastes could not be diluted or treated before testing for the characteristics.* But, in 1980, this approach apparently seemed unenforceable. Nonetheless, EPA has resorted to such a rule (40 CFR 268.3) with respect to land disposal restrictions (mid-1980s) proving that it is viable.

It is worth bringing up a semantic point that has plagued this field since its inception. Namely, what do we mean when we say "waste stream." Originally, the waste stream may have meant the *"point of generation"* of the waste, but early on, the term was adapted to mean the waste *"substance generated." Thus, the waste substance was imbued with a non-chemical intrinsic property of being "listed."* Unfortunately, *it takes political and legal alchemy to remove the stigma of "listing" from materials;* chemists are of no help.

Thus, with good intentions and poor science, the *listed hazardous wastes* were born. But, unlike the situation with characteristically hazardous wastes, if the waste is listed (or trapped in the listed category by certain rules discussed below), you cannot get out of the hazardous waste management system without a formal petition (40 CFR 261.20) and lengthy rulemaking (40 CFR 261.3(c)-(d)(2) and 260.22) called "delisting."

Remember the key to the rationale for a system of listed hazardous waste was that the operationally defined waste streams were specifically studied by the EPA. The burden of proof had been on the Agency to show through the rulemaking process that the operationally defined waste stream met the criteria set out by Congress. In principle, under this situation, the waste streams *at the point of generation* could be fairly certain to fail the existing characteristic test (or a characteristic test could be devised if it did not already exist); and through the listing process and the requirements for manifesting, these wastes would not be lost from the management system.

Starting in 1980, EPA proceeded to establish lists of waste streams from non-specific sources (e.g., still bottoms from recovery of various types of solvents, 40 CFR 261.31), waste streams from specific product-processes (e.g., recrystallization liquor from a specific type of reactor at a specific point in the manufacture of certain products by a certain process, 40 CFR 261.32) and a list of commercial products that become hazardous wastes when disposed (40 CFR 261.33).[338]

---

[338] There are a variety of features to these lists that are worth comment:

Remembering that the operationally defined waste streams had to be shown to meet the Congressional definition contained in RCRA, it is interesting that the operational definitions published in the list of waste from non-specific sources themselves often have concentrations in the definition. For example, several of these listed hazardous waste definitions (e.g., waste streams designated F001 - F005) read as follows, "The following ....[spent solvents]...containing, before use, a total of ten percent or more (by volume) of one or more...[solvents]...; and still bottoms from the recovery of these spent solvents and solvent mixtures." From these definitions, it is clear that the wastes that were being deemed to be <u>hazardous contained percent-levels of the hazardous constituents</u> and that concentration was critical to the argument used in the rulemaking for adding them to the list. The preambles to these listing rulemakings typically contain language as follows, "EPA has determined that these wastes contain **toxic constituents at concentrations** [emphasis added] which pose risks which are unacceptable....."

With regard to the identification of hazardous constituents and development of adequate test methods to monitor them, there is no conceptual reason that an exhaustive test protocol cannot be developed. Indeed, the work done by the EPA in developing the

The real problem arose when the *derived from, contained in* or *mixed with* rules were implemented. These rules were created from whole cloth by a branch chief in the EPA Office of Solid Waste (i.e., no Congressional mandate whatsoever). To prevent sham treatment, EPA took the drastic step of declaring that (i) when a hazardous waste (characteristic or listed) was treated to produce a new waste (that was *derived from* a hazardous waste), or (ii) when (non-hazardous) media was found *to contain* a (listed) hazardous waste, or (iii) when (non-hazardous) waste was *mixed with* a hazardous waste, then <u>the entire mass would be considered legally to be a hazardous waste (regardless of its characteristics)</u>. Thus, traces of a listed hazardous waste that were discovered in a contaminated medial (soil of water) generated in the course of investigating or remediating a contaminate site (e.g., under CERCLA see below) would be considered to be a hazardous waste. This problem arose (arises) on virtually every environmental remediation project and caused the cost

---

Hazardous Waste Identification Rules (published in 1995 and 1996) essentially establishes "bright line" concentration limits (i.e., "characteristics" by another name) for virtually every constituent of concern in waste streams and media. So, it is technically feasible to establish objective characteristics for all real hazards.

of remediation to soar. I have several interesting stories, but perhaps the most graphic was created by the USEPA itself.

Mount Dioxin

In this episode (1991- 1995), a USEPA project manager started excavating soil containing a listed hazardous waste from an industrial site near a residential neighborhood.

Aerial View of "Mt. Dioxin" Penscaola, FL circa 1997[339]

---

[339] This photo is sourced to http://www.cate.ws/mtd.html (Citizens Against Toxic Exposure) on the internet with no particular document. I suspect it was posted by an EPA employee. Black-and-white versions appear in various places on the internet.

After he had spent several million dollars creating a gigantic pile (i.e., the area of several football fields to a height of 50 feet, i.e., 344,250 tons of soil containing traces of dioxin), it was determined that disposal under the RCRA rules would cost upwards of a billion dollars to ship and burn the waste as required for dioxin-containing waste.

It happened to be an election year (1996) and when the neighbors complained, the Clinton-Gore Administration covered the pile with a tarp (about 2 million dollars) and moved the 358 families away (just in time for the election, October 18, 1996) at a cost of about 25 million dollars.[340] Mount Dioxin was maintained by the EPA essentially in this state until 2008, when the record of decision (ROD) was modified to allow onsite treatment including:

> Installing vertical and horizontal injection and extraction wells.

---

[340] This was the beginning of a concept called "Environmental Justice" i.e., reparations to minorities for living near industrial facilities.

> Treating contamination in source plume areas on site using chemical and bacterial methods designed to break down contaminants in soil and ground water.
>
> Treating contamination in high-level plume areas with bacterial methods designed to break down contaminants in soil and ground water.
>
> Using *monitored natural attenuation in dilute plume areas.*[341]
>
> Performing operation and maintenance activities.
>
> Placing institutional controls on the site to restrict ground water use.

Essentially, the site was returned to its original (pre-1990) condition with monitoring wells.

The Dioxin Dilemma

Biologists Kenneth Thimann (1904-1997) and Friedrich Went (1863 – 1935) discovered that plant growth and

---

[341] This translates into "do nothing" which is exactly what should have been done in 1991.

morphology is controlled by plant auxins (hormones) in 1937 with the isolation of indole-3-acetic acid:

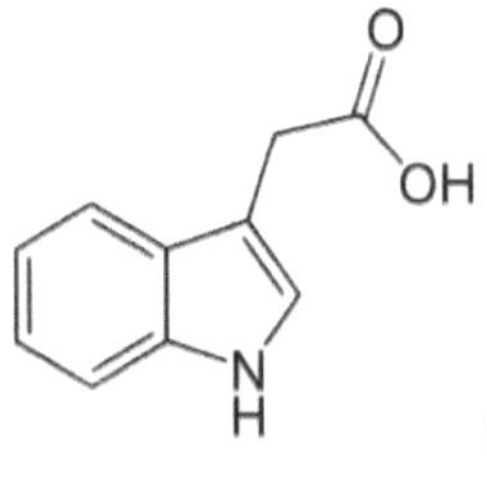

indole-3-acetic acid

Drawing by Yikrazuul, source Wikimedia Commons

Other naturally occurring auxins were soon discovered including 4-chloroindole-3-acetic acid and phenylacetic acid. It did not take long to discover that chlorophenoxyacetic acids were effective synthetic auxins. Since animals do not have receptors for these auxins, they are essentially non-toxic to humans.

Modern concern about "dioxin" (i.e., 2,3,7,8-tetrachlorodibenzo-p-dioxin) began with the manufacture of 2,4,5-T as a herbicide in the late 1940s:

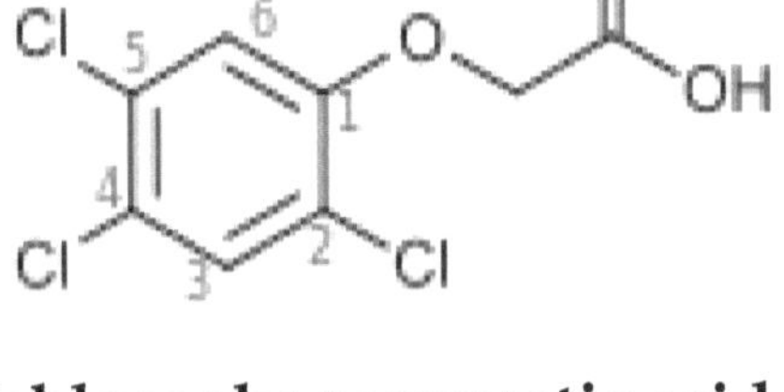

2,4,5-Trichlorophenoxyacetic acid (245-T)

Drawing by Monolemma, source Wikimedia Commons

Workers in these plants occasionally developed chloracne and over time this problem was traced to the presence of 2,3,7,8-tetrachlorodibenzodioxin. Toxicological bioassays with this compound discovered that certain strains of mice were incredibly sensitive to it; and in the atmosphere of the Rachael Carson era, it was quickly branded the most toxic substance known to man.

Into the 1970s, GC-MS had pushed routine analytical methods into the part per billion (10^{-9}) range, but the concern with dioxin prompted additional clean-up and pre-concentration steps that carried the sensitivity to the part per trillion range (10^{-12}). At this level of sensitivity, many unexpected compounds pop up, but only the dioxins and furans were monitored.

It is evident that "dioxin" arises in the production of 2,4,5-T by reaction of two moles of trichlorophenoxide with one another (nucleophilic aromatic displacements); and it became apparent that the "Agent Orange" defoliant used by US forces to eliminate ambush sites along trails and roads in Vietnam (1965-1970) was tainted with "dioxin."[342]

---

[342] This precipitated another issue concerning exposure of soldiers who applied the herbicide mainly by spraying from the

On July 10, 1976 a reactor overheated and burst a pressure control device at a herbicide manufacturing plant in Seveso, Italy releasing dioxin-laden mist over the town. All of this, of course, did not escape the attention of the Offices Toxic Substances and of Hazardous Waste at the USEPA.

The disposal of dioxin-containing waste became a matter of primary concern, with the levels of dioxin regulated in the *parts per trillion*. This is when various mass balance studies were conducted (measuring the amount of dioxin going into incinerators and the amount coming out of incinerators). The initial results suggested that dioxin was not destroyed in typical incinerators!

Note that typical waste incinerators of 1975 were designed to burn waste and then pass the exhaust gases through systems of air filters (bag- houses) at *high temperatures* that ensured (i) that the moisture (H_2O) and acid anhydrides (e.g., SO_2, NO_2) did not condense and cause corrosion and (ii) that the hot gases would rise rapidly up tall stacks for maximum dispersion.

---

air. I personally was exposed to 2,4,5-T in the 1960s using it to clear brush…with no noticeable effects.

The apparent failure to destroy dioxin in incinerators was a cause of alarm and was thus initially addressed by hotter furnace temperature and longer retention time. Oddly, this did not help and the carefully controlled data sets showed that *more dioxin was coming out of the incinerators than going in*. This was initially written off as experimental error (e.g., undetected sources of dioxin in the waste), but that argument was disproven. Thus, for a period of time in the 1980s, bureaucrats and environmental advocates sounded the alarm that *dioxin was indestructible*!

However, by the early 1990s, it was becoming obvious that trace levels of dioxin were being *produced* in the ordinary incineration processes. Among the environmental alarmists (inside and outside of EPA), there followed a general anti-chlorine chemical argument to *ban all chlorine-containing chemicals* with poly(vinylchloride) plastics (i.e., PVC) at the top of the list, followed by the pesticide pentachlorophenol.

There is nothing magical about dioxin. Like all organic compounds it is destroyed at about 1000°C. Thus, combustion engineers and chemists from the most threatened chemical industries discovered that the production of dioxins actually occurred *in the process of cooling the combustion gases (forming soot) from 1000°C to*

200°C. This range has now been narrowed to 600°C to 200°C. Above 600°C, dioxin is unstable (like most organic compounds) and below 200°C the reaction to form it is too slow (or is not thermodynamically favorable). The solution to the problem was to redesign incinerators to quickly quench the combustion gases from 1000°C to 100°C. The original design of the pollution control equipment turned incinerators into dioxin manufacturing systems.[343]

Again, the alarmists assumed that the mechanism of formation involved formation of chlorinated phenols followed by condensation to the corresponding dioxin. That's all the chemistry they knew. However, (unpublished) combustion trials with pressure treated wood by the USEPA at Research Triangle Park, revealed the interesting fact that combustion of pentachlorophenol-treated wood produced less dioxin than "chromated copper arsenate" (CCA)-treated wood. The reason for this was that the chlorination step (in the exhaust gases) is catalyzed by copper:

$$2 \; CuCl \leftrightarrows Cu + CuCl_2$$

---

[343] Now, backyard burn barrels used for disposal of household waste are considered the largest source of man-made dioxin release.

Apparently, Cu(I) chloride is a source of electrophilic chlorine under these conditions. The chlorine can come from anywhere (including salty breezes off the oceans). At the part-per-trillion level, lots of unusual chemistry is likely to occur. I suspect that the dioxin forming reaction involves addition of oxygen across the 9,10-positions of anthracenyl moieties (on soot) followed by C to O phenyl shifts to produce ortho-quinones, which then add across other anthracenyl moieties. This mechanism is not unlikely at the 10^{-12} level of yield.

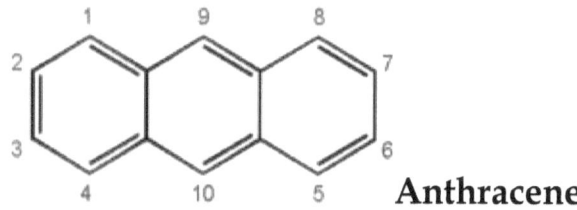 **Anthracene**

Drawing by Edgar181, source Wikimedia Commons

CERCLA

The Comprehensive Environmental Response, Compensation and Liability Act (CERCLA) had an amazing impact on the real estate industry in the late 1980s because it created "strict, joint and several," *retroactive* liability for land owners. The legal definition of "strict" means that the contamination could be completely unintentional and in keeping with normal (industry) standards of behavior. The term "joint and several" means that any individual or all individuals together could be held liable for contamination. And *retroactive* was merely inferred by an over-reaching agency.

It is worth addressing the *retroactive* (i.e., *ex post facto*) interpretation of CERCLA first to get it out of the way. Our Constitution clearly states that *ex post facto* laws at the State or Federal level are unconstitutional (Article 1 sections 9 and 10). The example I use with students these days is to say, "Suppose a change in the political winds brought Congress to the point of enacting laws against a commonly accepted practices such as abortion. What stops the implementation of laws that declare *abortion to be murder* from being applied to former practitioners of this procedure?" The answer is that the practitioners were aborting fetuses at a time when it was legal (not illegal); thus, under the Constitution the practitioners cannot be held liable for things they did when those things were legal. Similarly, most of the "legacy waste" (i.e., "releases of toxic substances to the environment" as defined in CERCLA) addressed under CERCLA was created *before* CERCLA was passed by Congress and before any specific regulations existed under RCRA and/or CERCLA.

Nonetheless, in a fashion I believe is typical of over-reach by the US Environmental Protection Agency, the implementation of the Federal law was initially applied indiscriminately. The "disconnect" here is that, under CERCLA, a specific tax was applied to certain chemicals in commerce to provide funds to clean up (legacy) hazardous waste sites and that tax was called the "Superfund," which was under the control of the USEPA. The EPA was expected to use this money itself (and through its contractors) to clean up legacy sites. But, the amount of money was so small compared to what the job became (i.e., what EPA deemed the job to be especially after SARA, 1986 and considering the RCRA rulings on contaminated soil and water, see above), that the EPA focused on a "polluter pays" philosophy through the 1980s taking the initiative of enforcement against many companies that did not understand the laws and were generally inclined to comply with EPA rather than risk the onerous (strict, joint and several) liability associated with the act. It is my impression that this sort of bullying (of private companies and other Federal agencies e.g., the Department of Defense and Department of Energy) by the EPA continued into the early 1990s when someone must have reminded the EPA Administrator of the Constitutional *ex post facto* prohibition, because she (Carol Browner (1955-)) started caveating her speeches with the date of CERCLA implementation. The point is that under CERCLA, EPA has the powers to investigate releases, characterize them and execute cleanup legacy waste *on private property*, but it (i.e., the actions for legacy waste) is all to be paid for out of the superfund, not applied as a liability to the land owner or company that disposed the waste (assuming that the release pre-dated the implementation of CERCLA).

The reason for the apparent resistance to CERCLA implementation (early 1990s) by industrial groups was complex, but it had a lot to do with the fact that up until that point (1980-1992), the administration of EPA had been in Republican hands and was not overly aggressive. With Browner and Gore (Vice President) managing the program, EPA was behaving much more hostilely towards anyone (land owners or industries) with the money to finance remediation projects. The other point was that, there was a fundamental change in CERCLA in 1986. Originally, CERCLA remedial response was to be driven by risk assessment, which involves assessing the actual risk (typically of cancer) experienced by real people and managing those risk.

Risk can be managed by removing the source of contamination or by simply creating a barrier between the source and the exposed population. In the eyes of most environmentalists (shared with "earth religionists") far too many sites were being managed in the early 1980s by paving them over and limiting the use of the land, ground water, etc. [see Mt. Dioxin above]. The environmental ethic demands "healing the earth," not putting a "bandage" on it.[344] Thus, in the mid-1980s, Congress passed the Superfund Amendments and Re-authorization Act (SARA). SARA forced risk management strategies to at least ensure that the remedial outcome left all the earth and water in compliance with all "applicable, or relevant and appropriate requirements" (ARARs)[345]. Thus, as part of the remedial planning, all laws, standards and regulations that *might be* ARARs had to be identified. The problem was that the EPA and cautious lawyers for risk-adverse "potentially responsible parties" (PRPs) uniformly behaved as though these were *all* at

---

[344] See for example the philosophy in *The Earth in the Balance: Healing the Global Environment* by Al Gore (1993).

[345] Pronounced A-raar-s.

least *"relevant and appropriate"* in defining remedial goals.

For example, the Safe Drinking Water Standards (SDWs) were promulgated for water utilities of significant minimal sizes (as supported by the Regulatory Impact Analysis (RIA) required for new regulations). Moreover, as a matter of practicality, no one would build a water supply system in a situation where water would likely be of insufficient quantity to meet demands; nor would a public water supply likely use very shallow water (percolating from the surface, directly into an unconfined aquifer) containing pathogens, natural levels of toxic materials or undesirable minerals (e.g., brackish water). Thus, the SDW standards would not be *applicable* to such an aquifer, because no one is actually drinking the water, and although the SDWA standards would be *relevant*, they would not be *appropriate* for such an aquifer. Thus, in most cases of contaminated ground water resulting from a small surface spill, the SDW standards sould not meet the criteria of an ARAR. But since that would potentially be determined in a court case involving a (small) business against the USEPA, the low-risk approach would be to not fight the EPA.

In any event, the trend should be to gradually work off the existing list of sites and the onerous liabilities of CERCLA continue to serve as a major deterrent to intentional releases. Of course, RCRA defines hazardous waste and proscribes how it should be managed (see above).

My biggest problem with the USEPA implementation of CERCLA is that they are exclusively concerned with toxicity and toxic hazard (e.g., cancer risk assessment, see above). These risks (e.g., in a CERCLA remediation) are the drivers of regulation and standards. However, toxic (cancer) risks are typically "hypothetical" in the sense that they *assume* a scenario that naïve people are exposed to the contaminant (through air, soil, water and food) for long periods of time[346] and (based on conservatively extrapolated dose-response data) may sustain disease or death (fatal cancer). If you weigh only these risks, you always come to the conclusion that less exposure is better.[347]

---

[346] Think of a post-apocalyptic family of four moving onto the site,

[347] You also satisfy the earth religionists' viewpoint that the earth must be healed.

But the facts are that environmental remediation projects are construction activities that involve *real statistical risk* of entirely different types. For example, remediation of a hazardous waste site not only involves risks to the on-site workers associated with chemical exposure, it also embodies all the risks of a conventional construction project including transportation accidents, falls, trench collapse, etc.

For example, we have excellent statistical data for traffic fatalities (based on real dead people not on a hypothetical model). However, EPA does not consider any of these risks (e.g., the transportation risk of moving Mt. Dioxin, see above) in its explicit analyses of remedial designs and I think this is a major mistake.

Green Chemistry

Sometime about the year 2000, I first heard the term "Green Chemistry." It reminded me of "Brown Fields." Brown fields was initially described as an EPA program to reduce the cost of environmental remediation by recognizing that some contaminated sites would be returned to industrial uses and thus really did not need to be returned to "green fields" conditions (i.e., by waving ARARs). By the time it passed through Congress in 1992 (Public Law 107-118 (H.R. 2869)), however, "brown fields" referred to lightly contaminated urban sites that were being taken to green fields conditions (with the aid of substantial grants from EPA):

> (C) …the President may authorize financial assistance under section 104(k) to an eligible entity at a site included in clause (i), (iv), (v), (vi), (viii), or (ix) of subparagraph (B) if the President finds that financial assistance will protect human health and the environment, and either promote economic development or enable the creation of, preservation of, or addition to parks, greenways, undeveloped property, other recreational property, or other property used for nonprofit purposes.

So, I had some doubts about what "green chemistry" actually meant. If some chemistry was deemed "green," was all other chemistry an affront to the environment?

Ultimately, I discovered that Green Chemistry was defined by the USEPA as follows:

> "Green chemistry is the design of chemical products and processes that reduce or eliminate the generation of hazardous substances. Green chemistry applies across the life cycle of a chemical product, including its design, manufacture, use, and ultimate disposal.

Green chemistry is also known as sustainable chemistry."

I failed (and still fail) to see what all the excitement was about circa 2000 because TSCA had clearly implied that idea in 1976. In fact, the projects I had worked on for the Office of Toxic Substances and Office of Solid Waste in the 1980s had implicitly or explicitly contained the idea of minimizing or eliminating both toxic substances and hazardous waste. There was a substantial economic incentive built into these laws to do that. The example of epichlorohydrin (discussed above) was clearly a specific effort by the EPA to evaluate alternative chemistry to eliminate from commerce a chemical that was believed to represent a hazard. Although we were not explicitly tasked (by OTS) to consider waste reduction, our experience working in both TSCA and RCRA (separate EPA contracts supported by the same consultant staff) told us that reduction of the volume and toxicity of the waste was a fundamental part of the cradle-to-grave philosophy of stewardship. So, we had included in our analysis estimates of the waste generation of different process options; and, of course, we considered cost (see table above). Similar, but less detailed, work was done on every TSCA PMN review and test rules development support document and OSW hazardous waste listing (adding waste to 40 CFR 261) support document.

It turned out that EPA had formally characterized "green chemistry" in 1991 through the efforts of Paul T. Anastas[348] (1962-) who appears to have been at the right place at the right time with the right political mindset. It is embodied in 12 guiding principles (source EPA):

> *1. Prevent waste: Design chemical syntheses to prevent waste. Leave no waste to treat or clean up.*
>
> *2. Maximize atom economy: Design syntheses so that the final product contains the maximum proportion of the starting materials. Waste few or no atoms.*
>
> *3. Design less hazardous chemical syntheses: Design syntheses to use and generate substances with little or no toxicity to either humans or the environment.*

---

[348] It is interesting that Dr. Anastas is frequently identified as "the Father of Green Chemistry," a title which he does not seem to try to squelch. And, this has created some consternation among admirers of Dr. John Warner (2014 Perkin Medal) who also is called the Father of Green Chemistry by his admirers. Their work was done about 15 years after TSCA was implemented, so I'm not sure how they fathered it.

4. Design safer chemicals and products: Design chemical products that are fully effective yet have little or no toxicity.[349]

5. Use safer solvents and reaction conditions: Avoid using solvents, separation agents, or other auxiliary chemicals. If you must use these chemicals, use safer ones.

6. Increase energy efficiency: Run chemical reactions at room temperature and pressure whenever possible.

7. Use renewable feedstocks: Use starting materials (also known as feedstocks) that are renewable rather than depletable. The source of renewable feedstocks is often agricultural products or the wastes of other processes; the source of depletable feedstocks is often fossil fuels (petroleum, natural gas, or coal) or mining operations.[350]

---

[349] There likely will be a trade-off of risks here. For example, you can build a balcony from wood with no antimicrobial preservatives; but that deck might fall with people standing on it. EPA seems to worry more about toxic risk than other hazards: transportation accidents, fires, falls, etc.

[350] This is obviously a dig at the petrochemical and coal industries. Ironically, plastic bottles are recently deemed a major problem although they are a tiny fraction of the fossil hydrocarbons recovered. If organic chemical products were the only use of fossil hydrocarbons, the supply would be essentially unlimited.

8. Avoid chemical derivatives: Avoid using blocking or protecting groups or any temporary modifications if possible. Derivatives use additional reagents and generate waste.

9. Use catalysts, not stoichiometric reagents: Minimize waste by using catalytic reactions. Catalysts are effective in small amounts and can carry out a single reaction many times. They are preferable to stoichiometric reagents, which are used in excess and carry out a reaction only once.

10. Design chemicals and products to degrade after use: Design chemical products to break down to innocuous substances after use so that they do not accumulate in the environment. [351]

11. Analyze in real time to prevent pollution: Include in-process, real-time monitoring and control during syntheses to minimize or eliminate the formation of byproducts.

12. Minimize the potential for accidents: Design chemicals and their physical forms (solid, liquid, or gas) to minimize the potential for chemical accidents

---

[351] I would add to this that a material would preferably be easily recycled (like aluminum cans) relative to acquiring virgin materials. And, I believe that use of disposed items as fuel ("heat recovery" if you prefer) is perfectly acceptable and often preferable to (expensive) recycling. As stated, the principle seems to anticipate wanton release to the environment.

including explosions, fires, and releases to the environment.

Although cost and profit are not mentioned in this list, it has been my opinion for some time that reduced cost and reduced environmental impact often go hand in hand. What chemical manufacturer would not want to have processes and products that follow these principles? Almost every listed principle conceptually reduces cost and/or liability (i.e., potential costs). However, there are interactions of these principles that may require sacrificing one to optimize another.

2. Organic Chemistry versus Carbon Allotropes and Molecular Biology

In the Post-Modern period, organic chemists are running out of "low hanging fruit." We seem to have a pretty good handle on basic reaction mechanism and the ability (in principle) to synthesize any molecule of interest. Fundamental breakthroughs are likely to be less and less common in organic chemistry *as it is normally defined*. In particular, I see the legacy of R.B. Woodward in total synthesis as being made obsolete by the advances in chromatographic and spectroscopic sciences. We can deduce the structure, optical activity and conformation of large complex molecules without synthesizing them. And, biological molecules (by definition) can be synthesized by biological systems (which we can modify with genetic engineering).

Organic Chemistry in Materials Science

Although I do not see much new action by organic chemist in the area of "materials science," clearly the polymer work of organic chemists has been very productive and is not yet exhausted. In particular, I think that organic chemistry can make major contributions to synthesis of pure carbon allotropes (e.g., nanotubes) and their functionalized deravatives. Although this field has been dominated by high-temperature reactions involving carbides (C_2^{2-}) or "volatile" polycarbon compounds (e.g., graphite), I think that organic chemists can come up with high efficiency, low temperature methods of forming poly-carbon compounds. Indeed, I have suggested[352] using 1,1-dichloroethene as a building block to produce polyacetylene or polyene (sp-hybridized) carbon

---

[352] G.E. Parris. *Patent pending. 15/927,407.*

polymers. The idea is obviously based on the decomposition of the plastic known as Saran™. I think there may be other options including the classical decomposition of sugar ($C_{12}H_{22}O_{11}$) with acid. To my knowledge, no one has ever characterized the carbon from this reaction. If it is amorphous, it might be a good starting material for nanotubes.

No doubt organic semiconductors, photoelectric cells and photosynthetic systems can be developed. But, most of these advances will be the product of large research projects, not single academic inventors.

The Organic Chemistry of Molecular Biology

Anytime you try to speculate about new products and processes, you are likely to come off looking like a fool. In contrast, we have before us a wide variety of very important organic compounds that organic chemists have essentially ignored. These are the compounds normally captured by the field of molecular biology. I can understand the reluctance of organic chemists to become active in this field. Prior to the mid-1990s, little was known of the structure or function of proteins, glycans, DNA, nucleosomes, tRNA/mRNA, ribozymes or non-coding RNA. And, organic chemists ("electron pushers" who like to see the structure and bond rearrangements in reactions) have a hard time with the rather sketchy descriptions provided by biologists.

Organic chemists are not satisfied with names like *factor, receptor, activator, binder* and are put off by their usage. But, I believe that organic chemists should embrace this science with a firm belief that *all of biology is based on chemistry*. If the biological system works, it is because the chemistry works.

My personal interest in molecular biology was spurred by the structure and operation of topoisomerase published by J.M. Berger et al. in 1996.[353] This is a little machine powered by changes in polarity caused by changes in phosphorylation and hydrogen bonding. All good organic chemistry concepts.

My view is that chemistry departments (especially the organic chemistry section) should align themselves academically with the biology department (especially the molecular biology area). Thus, I am surprised that in most/many universities the chemistry department continues to be associated with the physical sciences

---

[353] J.M. Berger et al. Structure and mechanism of DNA topoisomerase II. *Nature*. 379(6562):225-32 (1996).

while the molecular biology research is generally in an entirely different division of life sciences. I believe that if organic chemists became conversant in the concepts and problems of molecular biology (how to initiate transcription, how to activate genes, how to replicate DNA, etc.) they would find a range of problems that would be very profitable for their efforts. As it is, it appears that biologists are being expected to work out these problems on their own, literally *reinventing organic chemistry* along the way.[354]

---

[354] Thomas T. Tidwell (circa 2010) wrote and interesting outline "The First Century of Physical Organic Chemistry: A Prologue" which points out that in the 1960s, people though physical organic chemistry was "dead." As evidence of that premature funeral, he points to the Woodward-Hoffman rules (which were taught to me as an undergraduate in 1969). I believe this was the most recent fundamental contribution to physical organic chemistry, although some would argue that point. Tidwell ends his paper on a very upbeat summary of all the places that physical organic chemistry can be applied including:

"biochemical processes, the chemistry of disease, the design and understanding of molecular materials, the preservation of the environment on earth, and the understanding of chemistry in the Solar System."

Ironically, except for the punitive "green chemistry," these are not areas that organic chemists are being encouraged to follow

I also have a fundamental complaint about the teaching of general chemistry. It turns out (of course) that in most schools the students taking the introductory chemistry courses are mainly engineers (not chemists or biologists). Thus, in the presentation of topics like thermodynamics, the concepts are typically presented in a fashion more favorable to engineers than to "molecular scientists." I see much more similarity in biology and chemistry that I do with chemistry and (e.g.) mechanical engineering or even nuclear physics. Thus, if universities were organized more around a "molecular sciences" discipline and an ""engineering" discipline, I believe that course work would be more relevant to the tasks at hand.

3. The Career of the Organic Chemist

As discussed above, the days of entrepreneurship by chemists (especially organic chemists) are gone...unless you want to break the law and end up in jail. The burden of regulations placed on chemists today would

---

by the ACS or academic programs. These are the areas that "molecular biologists" are now grappling with.

ensure that Perkin, Nobel, von Heyden, Baekeland and others would more likely be in jail than at the Nobel Prize ceremonies.[355]

In my career, I have never been able to find a job in a research oriented setting with which I could pay my bills. Like most chemists, I have had to work outside my discipline. I have worked as an environmental consultant, where marketing and management skills were more important than my technical training; I have worked as a lobbyist, where interpersonal skills and legal knowledge were more important than my technical knowledge; I have sold automobiles for a living; I delivered UPS packages for two weeks; I worked in a production laboratory for 9 months; etc. But, since I left my post-doctoral position, I have never

---

[355] Actually, they would be sued first and then put in jail. Modern chemists are generally required as a condition of hire to agree to a variety of terms: (1) ownership agreements (i.e., the company owns anything the chemist discovers), (2) secrecy agreements (i.e., the chemists can only publish what the company believed to be in its best interest) and (3) non-compete agreements (i.e., the chemist cannot go out on their own and compete with their employer). Patents may list the chemist as the author, but they will be *assigned* to the company or the university or the government.

been in a position be paid to conceive of chemical experiments and execute them.

What I have learned is that anyone can call themselves a chemist. There is no professional distinction recognized in law. If you go into production laboratories (e.g., analytical labs doing routine analytical work) you will likely find (as I found) few people with actual degrees in chemistry. For the most part, they have other degrees and have been given training by various instrument manufacturers on how to operate the instrument they are running. In many states, there is some obscure law that a person who manages an analytical laboratory must hold a PhD in chemistry. But, that is hardly enforced. In the laboratory where I worked, the manager had business cards that said "PhD, ABD." I asked what the "ABD" stood for, and he told me "all but dissertation." Since he had been out of graduate school for at least 10 years, it was not a matter of him trying to finish writing his dissertation; he did not have enough work to write a dissertation. In other words, he was not actually a PhD, and had misrepresented himself for decades, and no one really cared.

We live in a world with professional engineers (PE), professional geologist (PG), physicians (MD), lawyers

(JD), certified industrial hygienists (CIH), certified underground tank installers, certified public accountants (CPA), licensed automobile salespeople, registered nurses (RN), licensed real estate appraiser, certified master automobile technicians, etc.; and any of these people can legally call themselves "chemist" if they wish. I once raised this issue with someone on the phone at the American Chemical Society (who ask me repeatedly to join their ranks) and they said, "Well, if you went to a school with an ACS certified program, you are a certified chemists." But the only people who seem to know about or care about that is the ACS itself. I've never had an employer ask if my program was ACS certified. Even within the academic community the only thing that people actually care about is your publications (how many and where). While I was a post-doc at NBS/NIST, my office-mate was also a post-doc, but it turned out he never got a BS or any sort of a bachelor's degree (he was kicked out of school for some prank).

The point of all this is that the profession of chemistry is strongly associated with a career in academia and only a lucky few will ever get to the level of a university with laboratory research options or an industrial research lab. A second group of chemists with advanced degrees will find jobs in non-research academia (e.g., 2-year programs) where they will spend the next thirty years of their lives teaching introductory general and organic chemistry. The rest of the people receiving degrees in chemistry, especially advanced degrees will end up working outside of their profession and training. For example, one good friend of mine in graduate school (with publications in organometallic chemistry and metal hydrides) worked his entire career in a textile firm; another went to medical school and ended up doing cleft palate surgery. I ended up in consulting: overseeing the field work of drillers and surveyors one week and sparing with bureaucrats the next.

But, it was not always this way. I and my contemporaries are the victims of various predictable forces that affect labor markets. We are actually doubly victimized because the people who we most trusted to advise us (our college mentors) were so poorly informed. Throughout my college career, I was a leading student and was given every encouragement to push ahead in my studies and my research. My mentors did not know that they were training me for a job that would not exist. Most of my professors (as I now realize) were only 10 to 20 years older than I was.

They certainly were not going to be giving up their jobs anytime soon. (I think the last one just retired…I'm 72 years old.) But they did not appreciate that they had been extremely lucky to have been born a few years ahead of a baby boon *and* a technological boom. Not only were they in a relatively small cohort (born during the depression or WWII), they were followed by the post-war baby boom and the technical demands of WWII, the cold war, Korea, and the space race where academic training was suddenly viewed as incredibly important and a host of new technologies were available for exploitation. As a result, most of them had had no trouble obtaining academic jobs (if they wanted them) or interesting jobs in industry. So, they were quite unaware of the problems the baby-boom cohort would face.

Of course, every professor holding down a research position was producing a new PhD (and probably more) ever year or two. And this, of course, meant immediate competition for new jobs. It was perhaps an alert to me that my government saw fit to draft me out of graduate school (23 years old and married) for the purpose of fighting the Vietnam War or at least supporting the national defense for several years. In contrast, during WWII, academics were pulled into exciting technical jobs. It turns out that I received my PhD in 1974 along with numerous other PhD organic chemists from the post war baby-boom:

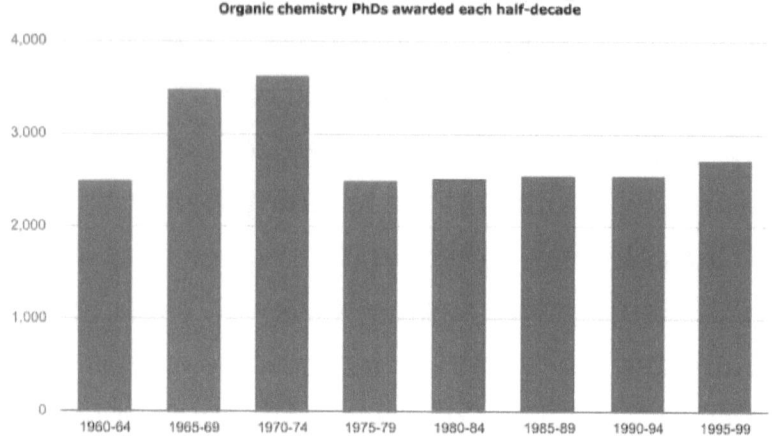

Organic chemistry PhDs awarded each half-decade

http://chemjobber.blogspot.com/2010/12/chart-of-day-organic-chemistry-phds.html

Ironically there seems to be a continuous over-supply of chemists in the US. The following post form the chemjobber.blogspot.com (2010) is completely consistent with my experience:

Anonymous December 26, 2010 at 9:42 PM

I am one of those organic chemistry PhD chemists from the early 70s. The factors for so many of us chemists from that era were:

1) the baby boom #s

2) we were inspired to be scientists by society's worship of atomic, space, agrichemical and pharmaceutical scientists' betterment of all our lives and their role in winning WWII. Remember plastics!![356] and nylon stockings and DDT?

3) we could play with chemicals in back yard laboratories making all kinds of obnoxious things like bromine, white phosphorus, nitric acid, lead azide, silver fulminate and nitroglycerin without worrying about the fire marshal, DEA, AFT or EPA taking us to jail or showing up in bunny suits. Yes I made them all by age 14. You could buy all kinds of chemicals and glassware at the local hobby shop back then. I even purchase sodium azide, and it was sent to me by US mail!

4) we were highly employable and highly valued by industrial employers. Chemists today have no idea how well scientists were treated by corporate types in the 50s and 60s. We were considered company crown jewels.[357]

5) a vast expansion of academic employment and government funds for training the baby boomers

What changed?

---

[356] This is a reference to the movie *The Graduate* (1967) with Dustin Hoffman. If you have not seen it, you must.

[357] At least these were our assumptions as undergraduates.

1) The 1973 recession - see Nov 1971 Life Magazine photo of 200 job rejection letters stapled to a PhD chemists' lab ceiling.
2) Love Canal and the rise of chemophobia
3) The end of the baby boom
4) The increasing poor treatment of chemists by employers including lay-offs, downsizings and eroding pay vis a vis inflation.
5) The end of rapid expansion of university hiring and lower government funding
6) Better/cleaner/safer/lucrative opportunities than chemistry - biological sciences, computer sciences and quantitative finance
7) Diminished commitment of industry to chemistry sciences
8) There were so many of us we lost our value to society and industry

I would point out that the slow-down in employment actually happened in the late 1960s, but by then many of us were already in the pipeline and followed through with our plans to get a PhD, which we did by 1974. The dismal job market for PhD chemist in 1970-71 (*"Nov 1971 Life Magazine photo of 200 job rejection letters stapled to a PhD chemists' lab ceiling"*)[358] got undergraduates out of the chemistry PhD programs and into other disciplines.

While the jobs for PhD chemists has recovered and stabilized somewhat in the range of about 2500 chemistry PhDs per year:

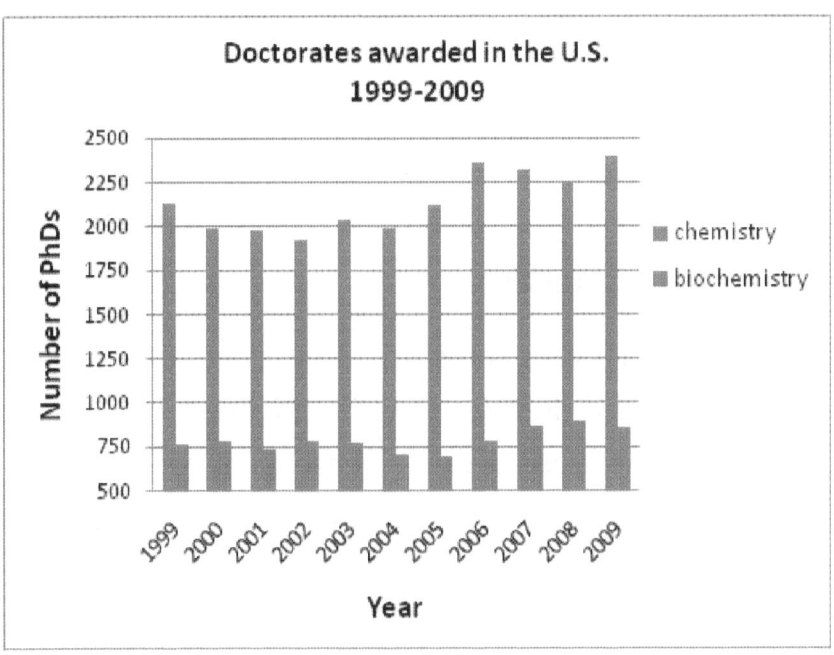

Graph by Leigh Krietsch Boerner[359] of data from NSF

---

[358] I well remember this full page photograph, but I cannot verify the month or year. In November 1971, I was in the Army at Ft Bragg, NC.

[359] http://cenblog.org/just-another-electron-pusher/2010/12/too-many-phds-thats-anybodys-guess/

But, I must point out that while the overall market for PhDs in chemistry has stabilized somewhat, the makeup of those hired has changed drastically. Basically, in one generation (1970-2010) the academic PhDs went from essentially *all white males (most American-born)* to about *25% women and as much as 50% international*. The large international contingent rarely represents desirable import of novel ideas from foreign countries (i.e., talent that is not available in the US) but more often the H1B visa are merely opening the US job market to foreign students.

Although the H1B visa is technically for 3 to 6 years, it may be extended indefinitely while the holder applies for permanent resident status. There is, of course, a worker protection clause in the law; but it is almost impossible for an individual to realize that he or she has been displaced by a foreigner and even if he/she suspects that the employer has not provided the higher pay etc., it is not practical for an individual (who is presumably unemployed) to attempt to fight against an employer (with whom he/she seeks employment). Thus while females and minorities (including foreign-born immigrants) have seen an improvement in employment options over the last 30 years, the academic experience for "(old) white men" has generally been dismal.

www.ingramcontent.com/pod-product-compliance
Lightning Source LLC
Chambersburg PA
CBHW021809170526
45157CB00007B/2514